Sharing Knowledge, Shaping Europe

Transformations: Studies in the History of Science and Technology

Jed Z. Buchwald, general editor

Dolores L. Augustine, *Red Prometheus: Engineering and Dictatorship in East Germany, 1945–1990*

Lawrence Badash, *A Nuclear Winter's Tale: Science and Politics in the 1980s*

Lino Camprubí, *Engineers and the Making of the Francoist Regime*

Mordechai Feingold, editor, *Jesuit Science and the Republic of Letters*

Larrie D. Ferreiro, *Ships and Science: The Birth of Naval Architecture in the Scientific Revolution, 1600–1800*

Gabriel Finkelstein, *Emil du Bois-Reymond: Neuroscience, Self, and Society in Nineteenth-Century Germany*

Kostas Gavroglu and Ana Isabel da Silva Araújo Simões, *Neither Physics nor Chemistry: A History of Quantum Chemistry*

Sander Gliboff, *H. G. Bronn, Ernst Haeckel, and the Origins of German Darwinism: A Study in Translation and Transformation*

Niccolò Guicciardini, *Isaac Newton on Mathematical Certainty and Method*

Kristine Harper, *Weather by the Numbers: The Genesis of Modern Meteorology*

Sungook Hong, *Wireless: From Marconi's Black-Box to the Audion*

Jeff Horn, *The Path Not Taken: French Industrialization in the Age of Revolution, 1750–1830*

Alexandra Hui, *The Psychophysical Ear: Musical Experiments, Experimental Sounds, 1840–1910*

Myles W. Jackson, *The Genealogy of a Gene: Patents, HIV/AIDS, and Race*

Myles W. Jackson, *Harmonious Triads: Physicists, Musicians, and Instrument Makers in Nineteenth-Century Germany*

Myles W. Jackson, *Spectrum of Belief: Joseph von Fraunhofer and the Craft of Precision Optics*

Paul R. Josephson, *Lenin's Laureate: Zhores Alferov's Life in Communist Science*

Mi Gyung Kim, *Affinity, That Elusive Dream: A Genealogy of the Chemical Revolution*

Ursula Klein and Wolfgang Lefèvre, *Materials in Eighteenth-Century Science: A Historical Ontology*

John Krige, *American Hegemony and the Postwar Reconstruction of Science in Europe*

John Krige, *Sharing Knowledge, Shaping Europe: US Technological Collaboration and Nonproliferation*

Janis Langins, *Conserving the Enlightenment: French Military Engineering from Vauban to the Revolution*

Wolfgang Lefèvre, editor, *Picturing Machines 1400–1700*

Staffan Müller-Wille and Hans-Jörg Rheinberger, editors, *Heredity Produced: At the Crossroads of Biology, Politics, and Culture, 1500–1870*

Staffan Müller-Wille and Christina Brandt, editors, *Heredity Explored: Between Public Domain and Experimental Science, 1850–1930*

William R. Newman and Anthony Grafton, editors, *Secrets of Nature: Astrology and Alchemy in Early Modern Europe*

Naomi Oreskes and John Krige, editors, *Science and Technology in the Global Cold War*

Gianna Pomata and Nancy G. Siraisi, editors, *Historia: Empiricism and Erudition in Early Modern Europe*

Alan J. Rocke, *Nationalizing Science: Adolphe Wurtz and the Battle for French Chemistry*

George Saliba, *Islamic Science and the Making of the European Renaissance*

Suman Seth, *Crafting the Quantum: Arnold Sommerfeld and the Practice of Theory, 1890–1926.*

William Thomas, *Rational Action: The Sciences of Policy in Britain and America, 1940–1960*

Leslie Tomory, *Progressive Enlightenment: The Origins of the Gaslight Industry 1780–1820*

Nicolás Wey Gómez, *The Tropics of Empire: Why Columbus Sailed South to the Indies*

Sharing Knowledge, Shaping Europe

US Technological Collaboration and Nonproliferation

John Krige

The MIT Press
Cambridge, Massachusetts
London, England

Set in Sabon by the MIT Press. Printed and bound in the United States of America.

Library of Congress Cataloging-in-Publication Data

Names: Krige, John, author.
Title: Sharing knowledge, shaping Europe : U.S. technological collaboration and nonproliferation / John Krige.
Description: Cambridge, MA : MIT Press, [2016] | Series: Transformations : studies in the history of science and technology | Includes bibliographical references and index.
Identifiers: LCCN 2016001875 | ISBN 9780262034777 (hardcover : alk. paper)
Subjects: LCSH: Nuclear nonproliferation--Government policy--United States. | Nuclear nonproliferation--Government policy--European Union countries. | Nuclear industry--European Union countries. | Technology--International cooperation--History--20th century. | Euratom. | United States--Foreign relations--European Union countries. | European Union countries--Foreign relations--United States.

Classification: LCC JZ5675 .K75 2016 | DDC 327.1/747094--dc23 LC record available at http://lccn.loc.gov/2016001875

10 9 8 7 6 5 4 3 2 1

Contents

Acknowledgments

This book has been on my mind for almost a decade. I first encountered the use of technological collaboration as a tool of US foreign policy in my work on Eisenhower's Atoms for Peace program and on NASA's efforts at international collaboration. On the face of it this is a straightforward idea. Working out its details in practice was challenging if only because it was always embedded in a cluster of other foreign-policy concerns. The dovetailing of technological collaboration with the promotion of European integration and the control of nuclear weapons and ballistic missile technologies came to define my problematic, both opening doors and setting limits on the scope of what I wrote.

These ideas have been developed in many talks and in some published papers. I thank all those, too numerous to mention, who contributed to my thinking as it evolved. However, this book goes beyond what I have said or written before, enriching and improving my analyses. Two conferences were particularly useful in this regard. The first was organized by Bernd Greiner at the Hamburg Institute for Social Research. The second was organized by Roland Popp and Andreas Wenger at the ETH in Zurich. Those two gatherings exposed me to new thinking in Cold War history and security/non-proliferation studies respectively, enabling me to better contextualize the United States' technological collaboration with its European allies in the 1950s and the 1960s.

Some people deserve special mention. I have made extensive use of newly declassified nuclear history documents that William Burr regularly posts on the Internet accompanied by excellent commentaries. Bill also gave me a good deal of help personally, for which I am most grateful. Two of my colleagues at Georgia Tech, Dan Amsterdam and Jenny Smith, provided excellent feedback on several chapters. Mark Kramer was always available to clarify details, drawing on his encyclopedic knowledge of the literature on Cold War history. I relied often on Alex Wellerstein's expertise and his wonderful nuclear secrecy blog. Jake Hamblin made just the right comments as the manuscript was nearing completion.

This book was mostly written during a sabbatical break as the Liu Institute Visiting Fellow in Residence at Green College, University of British Columbia, Vancouver, Canada. My sincere thanks to Moura Quayle and Julie Wagemakers (Liu Institute) and to Mark Vesey (Green College) for their generous hospitality, amplified by the wonderful staff at the Liu and by the extraordinary group of graduate students in residence at Green College. My family stood by me as they always do. Lydie's help, support and enthusiasm nourished me throughout. I thank them all.

Primary Sources

Archives

COLE Sterling Cole Papers, Special Collections and University Archives, Colgate University Libraries, Hamilton, New York

HAEU Historical Archives of the European Union, Florence

LBJL Lyndon Baines Johnson Presidential Library, Austin, Texas

NARA National Archives and Records Administration, College Park, Maryland

NASA National Aeronautics and Space Administration Historical Reference Collection, Washington

UKNA The National Archives, Kew Gardens, London

Online Sources

NSA385 Early Atomic Energy Commission Studies Show Concern over Gas Centrifuge Proliferation Risk, National Security Archive Electronic Briefing Book No. 385, available at http://www2.gwu.edu/~nsarchiv/nukevault/ebb385

NSA518 William Burr, The Gas Centrifuge Secret: Origins of a US Policy of Nuclear Denial, 1954–1960, National Security Archive Electronic Briefing Book No. 518, available at http://nsarchive.gwu.edu/nukevault/ebb518-the-gas-centrifuge-secret-origins-of-US-policy-of-nuclear-denial-1954-1960/index.html

WCDA Wilson Center Digital Archive, Woodrow Wilson Center—particularly William Burr, US Secret Assistance to the French Nuclear Program, 1969–1975: From "Fourth Country" to Strategic Partner, available at https://www.wilsoncenter.org/publication/us-secret-assistance-to-the-french-nuclear-program-1969-1975-fourth-country-to-strategic

Introduction

The Stakes

Would the United States risk New York to save Paris? President Charles de Gaulle of France suspected that it would not. The two superpowers, he said in 1962, "were capable of … destroying each other," a risk that they were not likely to take for the defense of Europe.[1] France needed its own nuclear deterrent, if only to "tear the arm off the Russian bear." In the late 1950s, President Dwight D. Eisenhower had favored arming Europeans with nuclear weapons effectively under their control.[2] This liberal policy was anathema to Congress and had little support in the Department of State. As Secretary of State John Foster Dulles put it in May of 1956, it was "appalling to contemplate a multiplicity of uncontrolled national atomic developments leading to multiplying atomic weapons programs."[3] By 1960 the US government was firmly opposed to independent national nuclear forces. In an oft-quoted remark made in a 1963 "news conference," President John F. Kennedy insisted that, unless the US took a firm stand against proliferation, the number of nuclear powers might have grown to ten by 1970 and to fifteen or twenty by 1975 (the so-called *N*th-country problem). Kennedy's Secretary of Defense Robert McNamara emphasized that the US was committed to defending Europe with nuclear weapons. But doing so would make sense only in a US-led and "centrally controlled" campaign. Independent European deterrents such as that sought by de Gaulle were unnecessary and dangerous, and might easily destabilize the carefully planned, graduated US response to a Soviet threat.[4]

The US was particularly concerned about how France's acquisition of nuclear weapons might affect the Federal Republic of Germany (the

FRG; West Germany). Indeed, the "German question" plagued US policy makers for twenty years after World War II. They were constantly prey to an abiding fear that, if not checked, German militarism and nationalism would once again drag the international system into another world war. In response, US Secretary of the Treasury Hans Morgenthau went so far as to suggest in 1944 that Germany be converted into a country "primarily agricultural and pastoral in its character."[5] The counter-argument prevailed as the fault lines of the Cold War emerged. A democratized Germany's industrial might should be rebuilt and integrated into that of the Free World. The aim of American policy became the "double containment" of Soviet communism and of German nationalism. To combat these dangers in the 1950s, senior American officials, backed by their president, worked along with influential fractions of elites in major Continental countries to co-produce a supranational "United States of Europe" that officially came into being after the ratification of the Treaties of Rome in January of 1958. An integrated "Europe of the Six" (France, Germany, Italy, Belgium, the Netherlands, and Luxembourg) would enjoy peace through closer economic and political integration, and strength through the modernization of its science, its technology, and its industry. Security guarantees would be provided by the North Atlantic Treaty Organization (NATO) under American command. Integration would dilute Germany's sovereignty and keep its aspirations to great-power status in check.

De Gaulle's return to power in 1958 added a new uncertainty to the equation. On the one hand, he was welcomed by the US Department of State as providing France with much-needed stability as it struggled to define its role in Europe and to deal with its colonial war in Algeria. On the other, de Gaulle's emphasis on re-establishing France as a great power and his resistance to any institutional entanglements that constrained French sovereignty threatened to strangle the movement for European integration at its birth, challenged the basic premises of the Atlantic alliance under American leadership, and placed an enormous strain on NATO (from which de Gaulle successively withdrew command over French forces). De Gaulle's determination to pursue an independent nuclear deterrent was particularly irksome. It might draw the United States into a nuclear war with the Soviet Union, and it risked producing resentment in Germany, which was denied direct access to nuclear weapons. Admittedly, in 1954, Chancellor Konrad Adenauer had publicly committed Germany to never producing nuclear weapons on its soil. However, many American officials felt that that commitment had only been made *rebus sic stantibus* (so long as conditions have not substantially changed), and that Adenauer favored West Germany's eventually having nuclear weapons of its own.[6] And who could predict the attitude of Adenauer's successors? Under Secretary of State George Ball remarked in 1963 that "nothing could be more dangerous for the peace of the world than Germany with a grievance and a sense of isolation."[7] While it might be tempting to share technology to secure its allegiance, a nuclear-armed FRG would threaten the Soviet Union and destabilize the fragile balance of power that was being constructed in Europe. In short, the US felt obliged to do what it could to curb the development of nuclear weapons systems in France, not only because de Gaulle was unpredictable, but also because they would find themselves pressured by the FRG for nuclear assistance—playing havoc with US-USSR relations at the front lines of the Cold War. As President Kennedy emphasized to British Prime Minister Harold Macmillan in May of 1961, "the fact that the Germans are not now tempted to join or imitate the French program is due, in no small part, to US opposition to *N*th-country programs and to the uncertain prospects of the French (or any other) program in the absence of US aid." Kennedy continued: "If we were now to provide aid to France, and thus signal a major reversal of our opposition to *N*th-country programs, the likelihood that the Germans would eventually wish to acquire a nuclear weapons capability would be significantly increased."[8]

What makes for a stable international system?[9] This question, which Marc Trachtenberg has suggested should guide studies of international relations, informs the argument in this book. Indeed, it is implicitly the question that exercised American policy makers as they grappled to contain German revanchism and Soviet communism. As war turned to a fragile and contested peace, many dreaded that nationalistic German aspirations would once more tear Europe apart, leaving it prostrate before a Soviet Union that was "animated by a fanatic faith, antithetical to our own" and that sought to impose its "absolute authority over the rest of world" using "increasingly terrifying weapons of mass destruction" if necessary.[10] Arguably only the United States had the resources and the political will to stabilize a postwar situation defined in such apocalyptic terms. It was driven by the ideological conviction that, as the historian

Michael Hogan puts it, the "leadership of the free world was a sacred mission thrust upon the American people by divine Providence and the laws of both nature and history."[11] Thus inspired, senior American officials worked along with influential fractions of local elites to co-produce a new integrated Europe that could contribute to the defense of the Free World under American leadership.

The US could not stop France or the FRG from acquiring nuclear weapons. However, it could take measures of many kinds to discourage either country from doing so. It provided security guarantees through NATO. It proposed to integrate the FRG into a seaborne multilateral nuclear force that, in Hal Brands' happy phrase, would put nuclear weapons within West Germany's reach but out of its grasp.[12] It actively promoted, along with the Soviet Union and the United Kingdom, the Nuclear Non-Proliferation Treaty, which sought to freeze the distinction between nuclear haves and have-nots at its January 1967 level, thereby keeping the FRG out of the nuclear club for as long as possible. And it embarked on various strategies directed at the technological core of dual-use (civilian and military) nuclear and space programs in Europe to reshape them as best it could. These technological strategies are the stuff of the case studies in this book.

The Argument

This book describes how the United States used its technological pre-eminence as an instrument of "soft power" to direct the trajectories of nuclear and space/missile programs in Western Europe along paths that fostered integration, promoted non-proliferation, and defused challenges to the United States' technological superiority. Two technical agencies, the Atomic Energy Commission (AEC) and the National Aeronautics and Space Administration (NASA), were central to this strategy. The Department of State, other branches of the administration, and the president himself worked along with the AEC and with NASA to exploit technological opportunities that served foreign-policy goals. By promoting the civilian side of "dual-use" nuclear and space technologies in an integrated political framework, they hoped to stop or at least retard other countries' military programs, to narrow the trans-Atlantic technological gap, to modernize the economic base of Western Europe, and to open new markets for American firms.

The term "soft power" was coined by Joseph Nye Jr., a distinguished scholar of international relations at Harvard University's John F. Kennedy School of Government who has served as an adviser to several US administrations. For Nye, military force and economic incentives are not the only or indeed necessarily the best instruments for a dominant state to use to get others to do it what it wants. Though Nye's main concern is today's world his conceptual frame has wider application:

> Soft co-optive power is just as important as hard command power. If a state can make its power seem legitimate in the eyes of others, it will encounter less resistance to its wishes. If its culture and ideology are attractive others will more willingly follow. If it can establish international norms consistent with its society, it is less likely to have to change. If it can support institutions that make other states wish to channel or limit their activities in ways that the dominant state prefers, it may be spared the costly exercise of coercive hard power.[13]

Nye has emphasized the immense attraction that American science and technology have for people all over the world, far outstripping their admiration for anything else in the country, including its culture.[14] By highlighting the nexus of knowledge and power and by seeing its value as a foreign-policy tool that can shape priorities, norms and institutions "in ways that the dominant state prefers," he captures in a phrase the overall dynamic of the processes described in this book. In fact, individuals in successive US administrations dealing with nuclear and space policy imagined technological collaboration as an instrument of soft power even when they didn't use that term. It is an implicit actor's category.

This book does not treat the concept of soft power uncritically. On the contrary, it reveals its limits in sensitive domains. The use of soft power was undermined by disagreements within the Executive branch of the US government, by resistance from already-existing traditions and institutions, and by demands for national autonomy in Europe. Soft power is a useful organizing principle for this study. US officials discovered that it was often not a particularly effective policy tool in the cases discussed here, which posed particularly complex challenges to US foreign policy.

For some time now historians of science and technology, along with a growing number of diplomatic historians and historians of American foreign relations, have been exploring the role played by the transnational circulation of knowledge in consolidating American leadership of the Free World.[15] As the cultural historian Cristina Klein put it, revisionist historians now regard the period from 1945 to 1991 not only as an era marked by superpower rivalry but also "as a chapter in the ongoing process of globalization" during which "the Cold War was as much about creating an economically, politically, and militarily integrated 'free world' as it was about waging a war of attrition with the Soviets."[16] The pursuit of scientific and technological leadership did double duty in this fractured world. It was spurred by competition and rivalry with the Soviet Union in the first twenty years of the Cold War (though there was also some cooperation between the superpowers in the 1960s—for example, in global weather prediction and smallpox control).[17] It also provided a resource to be mobilized in collaborative technological projects with friendly countries, a bargaining chip that American officials hoped to use as a tool to fashion the technological choices of their European allies. In the event, the extent of American "leadership" and the permissible scope of scientific and technological collaboration turned out to be rather more restricted than had been hoped, hemmed in by concerns over technology transfer and national and economic security that gradually drained "soft power" of its content. Notwithstanding its limited success, the variety of creative strategies deployed in the nuclear and space fields remains an unusual and under-studied dimension of American foreign policy. It stands as a tribute to the faith that American officials placed in technology and technological leadership to resolve what were essentially political problems.

Technological Collaboration and Non-Proliferation

Nuclear and space technologies were highly sought after by many nations after World War II. Both were seen by aspirant global powers as defining markers of modernity and as essential to national strength and national prestige.[18] They were complementary from an offensive point of view: the ability to deliver nuclear warheads to their targets with ballistic missiles in a few minutes placed immense power in the hands of governments and posed a dramatic threat to the stability of the international system. Both also promised major social benefits with important commercial implications. The exploitation of atomic energy for peaceful purposes was heralded in the 1950s as inaugurating a third industrial revolution that would produce "energy too cheap to meter." It also provided a source of radioisotopes for use in agriculture, in industry, in medicine, and even in ecology.[19] Powerful rockets having a family resemblance to ballistic missiles not only promised to reveal new scientific knowledge about the universe; they also orbited applications satellites for telecommunications, navigation, weather, earth observation, and so on, that had major effects on business and social life.

Major European powers felt obliged to engage in nuclear and space activities.[20] To make the best of limited resources, they pooled money and manpower in collaborative efforts that they developed along with their national programs, looking to the United States for scientific and technological assistance. This dependence on American technological sharing provided the US administration with an opportunity to selectively encourage the military programs of Britain (a close ally) or to actively retard the development of de Gaulle's *force de frappe*. It could also be used as a non-proliferation tool to steer America's allies toward civilian projects. Returning from the FRG after a short visit in January of 1966, Donald Hornig told President Lyndon B. Johnson: "I cannot emphasize too strongly the intellectual leadership which the pre-eminence of American science and industrial technology gives us in the world today. It can be used as a powerful political force at little extra cost."[21] Hornig, Johnson's Special Assistant for Science and Technology, had been abroad to evaluate the state of science and technology in the FRG and to look into the possibilities for cooperation between the US and West Germany in areas that would "distract" the Germans from their putative interest in nuclear weapons. He was heartened by what he had found. He had also learned something that senior officials in the Department of State, the AEC, and NASA had known for years: The postwar commitment of the United States to scientific and technological leadership had not only produced a powerful military-industrial complex to hold communism at bay; it had also given the AEC and NASA leverage with which to steer European technological choices in favor of the peaceful (and commercial) uses of atomic energy and of space.

The promotion of "the peaceful atom" and of civilian space collaboration were hallmarks of the Eisenhower administration and were central to the missions of the AEC and NASA. In 1953 Eisenhower announced his Atoms for Peace program to rapturous applause at the United Nations.[22] It was implemented by the AEC using bilateral agreements with friendly governments that included safeguards to keep nuclear knowledge, materials, and technology from being diverted from reactors into national weapons programs. The Space Act that established NASA in 1958 gave it a dual mission: to secure American leadership in space science and technology and to collaborate internationally.[23] Heartened by these initiatives, Europeans, eager to enter first the Atomic Age and then the Space Age, actively sought what the US could give so as to learn from the leader, to avoid duplicating what had been done elsewhere already, and above all to save time and money. The US was willing to cooperate so long as its national security and economic advantage were not jeopardized by the sharing of sensitive information and technology.

Technological Leadership and Political Leverage

The quest for American scientific and technological pre-eminence was already on the agenda before World War II ended. It became a priority as the Cold War gained momentum. Qualitative technological superiority based on nuclear weapons, which replaced manpower with firepower, was initially justified as the only way to hold back Soviet "hordes" in Europe without mass mobilization and militarization on the home front. The challenges posed in the 1950s by Soviet scientific and technological achievements, including space spectaculars, called forth a broader and deeper American response. State-driven innovation and the rapid transformation of new knowledge into useful devices were seen to be necessary conditions for maintaining America's vitality and for the successful projection of American power abroad. Before 1940 the federal government had limited its support for science mostly to agriculture. That changed dramatically during World War II. Cold War rivalry permanently institutionalized federal sponsorship of research and development which became an essential complement to industrial research laboratories and to the support provided by American philanthropic foundations, notably the Rockefeller Foundation and later the Ford Foundation.[24]

The financial trajectory of the federal government's support for research and development after World War II mirrored the global challenges that the United States faced.[25] Having fallen back slightly in the late 1940s, it surged in response to the first Soviet nuclear test in August of 1949 and the outbreak of the Korean War in June of 1950. Funds to support R&D sponsored (mostly) by the Department of Defense and the Atomic Energy Commission more than doubled to $1.3 billion in fiscal year 1951 and more than doubled again to $3.1 billion in fiscal year 1953. They increased by 30 percent in real terms over the next few years to produce new high-tech weapons systems, to expand scientific manpower, and to encourage the advancement of basic science. The panic surrounding the launch of Sputnik in 1957 gave them another major boost. Technological achievements, from warheads and missiles to RCA's Whirlpool kitchen became markers of the superiority of one system over the other in the "technological cold war."[26] By 1967, federal support for R&D had almost quadrupled to $15 billion, much of it for NASA's civilian space program. Aaron Friedberg puts it this way: "From the onset of the Cold War, top American decision makers tended to believe both that it was necessary for their country to seek a technological edge over the Soviet Union and its allies, and that such an edge could be achieved and maintained. These beliefs helped to keep technology at the forefront of American strategy and to sustain a massive four-decade flow of resources into research and development."[27]

The pursuit of scientific and technological leadership dovetailed with the determination to secure American political and ideological leadership of the Free World. Congress embraced this role reluctantly in the early postwar period. The fear of a communist takeover in Greece and the Prague coup of March 1948 swung the balance against those who balked at further "foreign entanglements." By 1950 the administration of President Harry Truman was committed to containing communism in Europe and Asia and was pouring billions of dollars into the Marshall Plan. The president, through Point IV of his 1949 inaugural address, had offered scientific and technical assistance for the "improvement and growth of underdeveloped areas."[28] In an age in which the power of the modern state increasingly rested on its scientific and technological capability, scientific and technological "leadership" was an essential complement to political, economic, and military leadership. This was the ideological node that tied together America's global ambitions and the aims of the national innovation system.[29] It became the most important argument for the government's and industries' ongoing support for R&D, a spontaneous ideology that became so widespread as to be invisible.

The circulation of "dual-use" nuclear and space science and technology between the United States and Europe in the 1950s and the 1960s did not occur on a level playing field. The asymmetry that derived from US scientific and technological pre-eminence provided American administrations with a "powerful political force" (Hornig) with which they hoped to shape both Europe's research priorities and the institutional forms in which knowledge was produced. This fusion of technological leadership with political leverage should not be taken for granted. For example, the British were the leading producers of civilian nuclear power in Europe in the mid 1950s and hoped to dominate the global commercial market. That did not encourage them to engage in cooperative projects with Continental partners. On the contrary, they backed away from close collaboration, fearing that they would lose their competitive advantage and so lose their markets, through being obliged to share cutting-edge technology with their partners. Britain, as Stuart Butler puts it, sought to "preserve prestige through independence."[30] The United States, by contrast, sought to use its technological advantage in collaborative projects to learn what it could from others, to get a foothold in potential markets, and to shape the research priorities of its allies. Confident of its ability to exploit any new idea more rapidly than its competitors, and of the entrepreneurial dynamism of its industry, the US didn't balk at collaboration but rather embraced it to achieve scientific, commercial, and political objectives. Britain interpreted technological leadership in commercial terms. The US used it to establish a global footprint.

In a previous study of American hegemony and the postwar reconstruction of science in Europe, I showed how the US sought to use its scientific leadership to achieve political goals.[31] I argued that the Rockefeller Foundation did not simply foster the injection of American scientific knowledge into war-ravaged France immediately after World War II. It sought to *reorient* French science along US lines—to encourage molecular biology at the expense of "Lysenkoism," to decentralize research and break the central power of Paris, to make French science more international, and to integrate its personnel more tightly into the Anglo-Saxon research community. Isidor I. Rabi, in consultation with the Department of State, did not simply offer US support for the establishment of a new multinational high-energy physics laboratory in Europe in 1950. He made that support conditional on the fledgling FRG's being a member state of CERN and so dictated the scope of the research resources. To satisfy Rabi, the laboratory could not be equipped with a research reactor, as had been planned by its European proponents when had they modeled their project on Brookhaven National Laboratory (in the US) and on the Atomic Energy Research Establishment at Harwell (in England). Allied laws still in force denied German scientists access to nuclear power technology. Letting the FRG in thus meant keeping reactors out. After Sputnik, the Ford Foundation, through the NATO Science Committee, did not simply seek to strengthen engineering education in Europe. It sought, unsuccessfully, to implant the MIT educational model—a model that deliberately linked teaching and research with socially relevant needs and with military enhancement—in Europe. All of these initiatives brought together actors on both sides of the Atlantic in an asymmetric field of knowledge and power in which European partners worked along with their American counterparts to rebuild the strength of science and technology on the Continent. US actors sought to wield their scientific advantage and financial resources to mold programs and institutions in line with their conception of how research should be organized in Europe, sometimes clashing with entrenched local cultures that refused to bend to their inducements. Scientific and technological leadership bestowed political leverage (but did not guarantee political success).

The preponderance of US scientific and technological knowledge/power in 1945 gradually diminished over the time period covered here. Before World War II, or at least into the early 1930s, Europe had led the world in many fields, notably physics, chemistry, and mathematics.[32] The flight of scientists from fascism and the destruction of laboratories and industries during the war left the scientific and technological infrastructure in tatters. The US surged ahead in many areas of research, especially in costly and complex projects involving major investments in technological platforms. When European governments and their technical advisers set research priorities after the war, they often had to pool financial and human resources if they were to stay in the research game. Several options—purely national programs, bilateral arrangements with select partners (notably the US), multinational collaborative efforts—were on the table, but not all could be pursued simultaneously for lack of money and manpower. Ten years after the US had used atomic weapons on Japan, Francis Perrin, the technical director of the French Commissariat à l'Énergie Atomique told Premier Pierre Mendès France that France could not hope to embark on a bomb project just yet, because too many French scientists were left-wingers and were hostile to equipping their nation with nuclear weapons.[33] British nuclear scientists and engineers were asked in 1957 if they wished to participate in a collaborative European atomic energy program. The United Kingdom Atomic Energy Authority declined. Its staff members were so overworked on the national civilian and military nuclear programs that such participation "would cause the national nuclear energy program to collapse," they claimed.[34] The first successful French experiments with sounding rockets that soared about 80 miles above the Sahara took place in March of 1960, three years after the USSR had launched Sputnik into outer space. To the dismay of the French press, they relied on engineers who had come to France from Germany after the war to develop and launch the small rockets.[35] Even Britain, which in the 1950s led Europe in nuclear energy, in bomb and missile development, and in the production of civilian jet aircraft, struggled to maintain its capacity across a broad front (withdrawing entirely from missile and rocket development in 1971, for example).[36] In short, in the postwar period there was a lag between the United States and Europe of at least ten years in major technological fields. This provided an opportunity for the US to rebuild and reshape the technological strength of the region under American leadership to meet the communist threat.

Much of the transnational traffic in science and technology is driven by scientists, engineers, and industrialists themselves; unless matters of deep national concern are involved, the government is expected to facilitate the circulation of knowledge across borders, not regulate or impede it (e.g., by imposing export controls or visa restrictions on foreign nationals). By contrast, the machinery of government was centrally engaged in the two fields of knowledge that concern us here: nuclear and space technology. Inter-state negotiations touched on the security and commercial interests of both parties and demanded attention from the highest levels of the administration. Conflicts of interest were endemic, and desired outcomes often had to be secured by hard bargaining. In the 1950s, American officials embarked on several rounds of "trading exercises" to undercut those in Italy who wanted to develop a self-sufficient nuclear power program, seeking to lock them into an American-led technological trajectory that Simone Turchetti has described as "de-developing" the Italian nuclear power project.[37] "I need chips," US Secretary of State Henry Kissinger barked as he sought to drain Iranian petrodollars in the early 1970s, eventually persuading the Shah to invest in a massive American nuclear power program for his country, as Jacob Hamblin explains.[38] Nor should it be assumed that these negotiations always produced the outcome that the dominant state wanted. On the contrary, the exercise of soft power is a contested process with unpredictable outcomes, as we shall see time and again in this book. Negotiations over collaboration in sensitive technologies involve drawing a line between what can be shared and what will be denied. Denial can frustrate a dominant state's negotiating partners, undercut the reach of its soft power, and encourage the potential collaborators to strike off on their own.

The United States' foreign policy in nuclear and space matters was made with one eye on the needs of American industry. American utility companies were engaged from the start in discussions between the Department of State and its European partners on developing a nuclear power program on the Continent. The determination to help American utility companies get a foothold in Europe ahead of British competitors was one of the driving forces of the collaborative effort. American aerospace firms favored technological collaboration in rocketry with European partners after France became the world's third space power in 1965, orbiting its own satellite with an indigenously built heavy launcher derived from its missile program. Where the US government led, American industry was never far behind.

A final word on terminology: In speaking of nuclear and of space technologies in this book, I am referring to artifacts that only have meaning in a "technological-system-in-the-making." It takes many years and multiple resources (e.g. money, manpower, industrial capacity, and political will) to bring a sophisticated technological system into being. It has its own history, trajectory, twists, and turns, and it is constantly being upgraded. It is not to be imagined as a distinct object, but rather as an interconnected array of components and subsystems that depend crucially on one another and on the environment in which they operate. Even at a relatively advanced stage of development, technological systems may be functional but their performance may not be optimal. This is why the strategies described in this book—"foot-dragging," retardation, diversion, subversion—can be effective. Technological collaboration or denial can shape a technological trajectory, reorient it, accelerate it, or disrupt it, depending on the developmental phase at which it is invoked. A technological system is a complex, fluid, evolving thing, and there is no one best way to bring it to fruition: in an important sense it is always a work in progress.

The sources used in this book deal predominantly with negotiations within and between the administrative arms of governments in which broad lines of policy are laid down. Once authorized, strategies are implemented by multiple actors, including industry (organized, for example, in the Nuclear Suppliers Group since the mid 1970s)[39] and research scientists and engineers (in universities and weapons laboratories).[40] Both are present to a varying if muted degree in this study if only because robust documentary evidence of their roles is difficult to obtain.

The Findings

The argument is developed in chapters that proceed chronologically. They explore attempts by the United States to curb the proliferation of nuclear weapons and of nuclear delivery systems through various multinational technological arrangements brokered with select national actors in organizations such as Euratom (for civilian nuclear power, chapters 1 and 2) and ELDO (for satellite launcher development, chapter 4). These are complemented by bilateral arrangements with the FRG in space science (chapter 3) and with Britain in developing gas centrifuges for uranium enrichment (chapter 5). Focusing on this agenda from the point of view of knowledge circulation throws a different light on the role of the Department of State, the Atomic Energy Commission, and NASA in fostering European integration and thwarting proliferation. Not only were those agencies active in Europe well into the 1960s; they were always alert to the possibility of exploiting their technological leadership as an instrument of foreign policy, sometimes in the face of considerable hostility at home and abroad. To their dismay, the mobilization of technology as an instrument of soft power was seriously circumscribed. Whether because of techno-utopian hubris that exaggerated its possibilities or because of an underestimation of the resistance to their plans at home and abroad, technology as an instrument of soft power often failed to achieve the desired results. Technological leadership was most effectively wielded as an instrument of soft power when it was embedded in a web of supportive, overlapping power structures that "hardened" it. This "hardening" enabled the AEC to gain access to the British prototype that was central to a European project to use gas centrifuges to enrich uranium, defusing its proliferation risks to their satisfaction. It also shaped relations with France in the early 1970s, when the Nixon administration sought to strengthen the offensive and defensive capabilities of the French nuclear-weapons system.

Technological collaboration undoubtedly has risks, but so do technological denial and associated strategies of "hard power." There is no one best way to halt the spread of nuclear weapons, or to bring them under control. This book argues that technological collaboration, sponsored by the governments of leading powers and implemented by industries and research scientists and engineers with actual or would-be proliferators, can play a major role in framing rational policies to stabilize the world system. The exercise of "soft power" through technological sharing by scientists and engineers remains an important instrument to keep lines of communication open in an increasingly divided and dangerous world.

1

The United States and the Promotion of Euratom, 1955–56: Integration as an Instrument of Nuclear Non-Proliferation

The role of Euratom (the European Atomic Energy Community) in the postwar reconstruction of Europe is ambiguous. For a while, in the mid 1950s, Euratom was seen by many on both sides of the Atlantic as the spearhead of a revitalized move to supranational integration. The formation of an atomic energy community seemed to be the next step on the path to a United States of Europe. Several factors fueled these expectations. There was the euphoria surrounding the exploitation of "the peaceful atom" as a limitless source of energy. American help was guaranteed. The Department of State and the Atomic Energy Commission (AEC), with strong support from President Eisenhower's Atoms for Peace initiative, were willing to provide know-how, fissionable material, technology, and training to non-communist countries that wanted to develop their own programs. Energy supplies could be liberated from the political whims of foreign powers that controlled the supply of fossil fuels and could be placed on a reliable footing. And, unlike a common market, the pooling of resources in a sector that was in the early stages of development on the Continent was thought unlikely to fall prey to national rivalry. These ambitions were formalized in the Treaties of Rome, signed on March 25, 1957 and quickly ratified thereafter. France, the Federal Republic of Germany, Italy, Belgium, the Netherlands, and Luxembourg ("the Six") committed themselves to building a regional European Economic Community and placed their faith in Euratom to foster the production of nuclear energy to power economic growth. Notwithstanding this flying start, within a few years Euratom had lost its allure. Its exciting program to provide 15 million kilowatts of nuclear power for the Continent had already collapsed. The organization lived on, but it had to redefine its priorities in order to survive.[1]

The copious literature on European integration pays little attention to Euratom. It deserves more attention than it usually gets.[2] For one thing, the United States played a major role in its formation. Indeed, the negotiations over its program are a strong counterexample to those who would play down American influence on the process of European integration, at least in the mid 1950s. Geir Lundestad rightly insists that Euratom "could not have been established without Washington's strong backing" against divisions between the Six and Britain's push for a looser cooperative arrangement under the auspices of the OEEC (Organization for European Economic Cooperation).[3] That backing expressed the Eisenhower administration's commitment to embedding France and Germany in an institutional arrangement that would both foster European integration and curb the "national nuclear weapons problem," as it was called at the time.[4]

Most histories of Euratom focus on the diplomatic negotiations surrounding its origins and establishment. The technological core of the organization is always mentioned, of course, but its significance for the actors is downplayed; it provides the backdrop to the action rather than being at the heart of the action itself. This chapter and the next move the technological platform to center stage, treating Euratom as a major experiment in scientific and technological collaboration between the United States and an integrating Europe.

Klaus Schwabe has suggested that by the mid 1950s the US had "lost much of its previous leverage and prestige" on the Continent, and that much of what we know of European integration today—the common market and the European Commission in Brussels—"was the result of European initiatives ... rather than American nudges."[5] This overlooks the leverage and prestige bestowed by the putative technological leadership of the US in the nuclear field. In the hands of Department of State and AEC officials, its scientific and technological capability in all things nuclear was an important aspect of soft power. It provided them with a political weapon to tactfully encourage, and then actively support, a supranational structure against rival national and multilateral options. They were convinced that atomic energy could provide a "bonding incentive" for the Six on which, together, they could build an organization that would pave the way to closer economic and political integration while also curbing the weapons pretensions of some of Euratom's member states.

These two themes—Euratom as a platform for European integration and for non-proliferation—dominate the historiography of the organization. Whereas the former was emphasized in the 1990s, the latter has become increasingly prominent, reflecting the widespread concern today over the proliferation of nuclear-weapons programs around the world. As matters stand there is no consensus on the role of Euratom as an institution that curbed proliferation.

Gunnar Skogmar puts the containment of Germany at the heart of his analysis. He concludes that the development of a supranational organization that was dedicated to civilian nuclear power made a crucial contribution to non-proliferation, if only because it included Germany, diluting its sovereignty and establishing mechanisms that curbed its ambitions to acquire nuclear weapons. Indeed, according to Skogmar "the non-proliferation goal in US policy was stronger [in the 1950s] than is generally appreciated in the literature," Euratom being the "strategic missing piece for arms control of Germany."[6]

Marc Trachtenberg, on the other hand, sees Euratom as an instrument deliberately intended by its promoters in the United States to create the foundations of an "independent integrated European nuclear capability." Everyone knew that an atomic energy program was a dual-use program, and that "the military implications of the project," if exploited, would enable the Europeans "to achieve atomic independence."[7] Of course, as was mentioned in the introduction, US Secretary of State John Foster Dulles told Federal Minister of Atomic Affairs Franz Josef Strauss in May of 1956 that "a multiplicity of uncontrolled national atomic developments leading to multiplying atomic weapons programs was appalling to contemplate."[8] But an integrated European program was quite another matter, and was favored by Dulles.

The dual-use character of Euratom has been given a further twist by Grégoire Mallard, who bases his case on a close analysis of the treaties agreed between the US and Euratom when it was established. Mallard concludes that the intentions of Euratom's "federalist" proponents in Washington and in Europe were both peaceful-economic and military-strategic.[9] While the military aspects were kept secret, the civilian, economic potential was broadcast widely. The ensuing "charade," the *mise en scène* of peaceful use, was intended to dupe Congress and the Joint Committee on Atomic Energy, which would not have agreed to US support for Euratom if the military ambitions of its European proponents were openly displayed.[10] For Skogmar, Euratom was an instrument of arms control. For Trachtenberg it paved the way to European atomic independence. For Mallard, its peaceful ambitions were a front intended to distract attention away from its military program.

Many scholars pay little or no attention to developments in Euratom after the Treaties of Rome were ratified. They are barely touched on in documents reproduced in the handy *Foreign Relations of the United States* (*FRUS*) series published by the Office of the Historian in the Department of State. By contrast, this chapter and the next, drawing extensively on Department of State documents not published in *FRUS*, and keeping an eye firmly fixed on the negotiations over the technological content of the project, emphasize that the production of civilian nuclear power in an "integrated" organization of the Six was always the dominant goal of Euratom's proponents. Non-proliferation goals were recalibrated in 1956, when it was agreed that France could build nuclear weapons and remain a member of Euratom. They faded into the background in 1957–58. Much scholarship effectively truncates the history of Euratom in March of 1957. One must follow its fortunes for a little longer (see chapter 2 below) to appreciate the changing meaning of Euratom for its main protagonists and to properly assess the different weights given to its three main functions: a platform for developing nuclear power, an instrument of integration, and a curb on weapons proliferation.

Atoms for Peace—from Denying to Sharing Knowledge

> Euratom, not the IAEA, stood as the original cornerstone of Eisenhower's Atom's for Peace policy.
> —Jack Holl, Official Historian, US Department of Energy[11]

The optimism invested in Euratom in the mid 1950s combined utopian enthusiasm for the promise of nuclear energy with a radical expansion of the scope for sharing scientific and technological know-how made possible by the US Atomic Energy Act of August 1954. These legal changes were part and parcel of Eisenhower's promotion of the "peaceful" uses of atomic energy in response to a gloomy prognosis of the effects of nuclear war between the US and the USSR. It is important to spell out these developments in some detail. The Atoms for Peace program had a major effect on the United States' relations with its allies in the field of nuclear energy, and on the decision to establish Euratom.

The loss of the United States' monopoly of nuclear weapons and the escalating arms race caused considerable concern to the Eisenhower administration in its early days. The Soviet Union's advances in nuclear-weapons technology in 1949 (when it tested its first fission bomb) and in 1953 (when it tested its first fusion bomb) raised the specter of World War III. It was feared that before long the USSR might engage in a massive first strike in the hope of gaining victory before the US could retaliate effectively. It rapidly became evident that no technological measures could fully protect the US from the horrors of a nuclear attack: as the president put it a few years later, if that were to happen, "there just [weren't] enough bulldozers to scrape the bodies off the street."[12] Arms-control agreements were needed to achieve a measure of international stability, and seemed possible after Stalin's death (in March of 1953). But how could the trust needed for successful arms-control agreements be built, in view of the suspicion that had dogged all such attempts since the US first proposed its Baruch Plan to the United Nations in 1946?[13] A panel chaired by J. Robert Oppenheimer suggested that part of the solution lay in being candid with the American people about the balance of military power, spelling out in broad outline the current state of the American arsenal and what was known about the Soviet Union's capabilities.[14] "Operation Candor" had its downside, however. It antagonized the AEC's chairman, Lewis Strauss, who believed that secrecy was the key to American security, and that Oppenheimer was not reliable as an advisor on nuclear policy. Operation Candor also did nothing to reassure the public. On the contrary, as one internal minute put it, "all that it contained was mortal Soviet attack followed by mortal US counterattack—in other words, bang-bang, no hope, no way out at the end."[15] This apocalyptic scenario suddenly became plausible when the USSR successfully tested a new thermonuclear-like weapon in August 1953. In October, Eisenhower decided not to disclose any details of America's stockpile of atomic weapons. Operation Candor was dead.[16]

While vacationing in Denver in the summer of 1953, Eisenhower had come up with the "vague and general germ" of an alternative idea.[17] It was given substance early in September by Robert Cutler, his special assistant for National Security Affairs.[18] The scheme was refined after several rounds of discussion in the administration and was announced to the UN General Assembly on December 8, 1953 in what came to be known as the Atoms for Peace Speech. Eisenhower shared a draft with Prime Minister Winston Churchill at a meeting in Bermuda shortly before he addressed the UN. Churchill was so impressed that he later told the British Parliament that it was "one of the most important events in world history since the end of the war," and that it "might prove to be a turning point in our destiny."[19]

The significance of Eisenhower's speech lay in its attempt to break the deadlock of mutually assured destruction—to refuse "the hopeless finality of a belief that two atomic colossi are doomed malevolently to eye each other indefinitely across a trembling world." What was needed, Eisenhower said, was to "move out of the dark chamber of horrors into the light," to "find the way by which the miraculous inventiveness of man shall not be dedicated to his death, but consecrated to his life." The solution he proposed was that the nuclear powers, notably the United States and the Soviet Union, should make joint contributions from their stockpiles of "normal uranium and fissionable materials" to an atomic "pool" that would be managed by an International Atomic Energy Agency (IAEA) set up under the auspices of the United Nations. Its main task would be to "devise methods whereby this fissionable material would be allocated to serve the peaceful pursuits of mankind"—for research in agriculture and in medicine, and to provide electrical energy in the "power-starved areas of the world."[20] Eisenhower's proposals were greeted with rapturous applause. When he sat down, his eyes were glistening with emotion.

Eisenhower liked this idea because it linked a small step toward disarmament with a big step toward sharing the benefits of atomic energy with the rest of the world. The call to divert nuclear material to an "atomic bank" put the Soviets on the spot: if they refused they would be seen as warmongers, and if they accepted they would sacrifice some of their precious enriched material from their bomb project. The use of that material to spread the benefits of the atom under international control would please foreign governments and their nuclear communities, who sought access to fissile material and to scientific and technological know-how. It would also create new markets for American power companies.

The "disarmament" aspect soon got bogged down in mutual suspicion between the three nuclear-weapons powers. One major sticking point was the amount of fissile material that each partner should contribute to the pool. Eisenhower's original idea, and the one that seems to have intrigued him when he was in Denver, was that the US and the USSR each "turn over to the United Nations, for peaceful use, X kilograms of fissionable material," where "the amount X could be fixed at a figure which we could handle from our stockpile, but which it would be difficult for the Soviets to match."[21] The Soviets had no intention of falling into that trap, and quickly pointed out that the scheme could work only if they and the US diverted so little material from their stockpiles as not to affect their weapons programs—i.e., so little that it was irrelevant as a disarmament measure.[22] Lord Cherwell, Churchill's chief scientific adviser, made the same point.[23]

In May of 1954 the Eisenhower administration changed tack and decided that the US would move ahead with its plans for the IAEA unilaterally, maintaining pressure on the UK and the USSR to make some material contribution to the atomic "bank." The priority now was to promote the peaceful uses of atomic energy more widely in anticipation of the IAEA being established.[24] The AEC would make fissile material available for use in research and power reactors through bilateral government-to-government agreements with friendly countries—agreements that would incorporate safeguards against diversion to military purposes. In addition, existing legislation would be revised to enable both domestic power companies and foreign partners to have access to enriched uranium and previously classified knowledge on nuclear science and technology.

America's allies that had nuclear aspirations were delighted with these developments, which were indicative of a radical step away from the philosophy that had inspired the postwar nuclear regime that the US had put in place. That regime was intended to secure the American monopoly of atomic weapons for as long as was possible. It was based on a policy of exclusion and denial, both as regards access to fissionable material, and as regards nuclear scientific and technological knowledge.[25] Strict constraints on the dissemination of scientific and technological information were enshrined in the Atomic Energy Act of 1946 (the McMahon Act). That act defined a new legal term: "restricted data," meaning data "born secret." Restricted data included all information regarding the manufacture and use of nuclear weapons as well as the production of fissile material or its use to produce nuclear power.[26] These provisions seriously hampered relationships with allies, and suffocated the development of a domestic nuclear energy industry. By 1952 they no longer served their purpose. The secret was out: both the Soviet Union and the United Kingdom had tested nuclear weapons, and both were embarking on civilian nuclear power programs.

The Atomic Energy Act of August 30, 1954 provided for the need to disseminate, rather than deny, nuclear information and fissionable materials, and to make them available both to domestic power companies and to the rest of the world. Not only did it allow the AEC to grant private entities and foreign countries access to certain classes of "restricted data"; it also increased the government's commitment to declassifying knowledge pertaining to nuclear reactors and to certain methods of enriching uranium. These changes did not go unchallenged. Some members of the Atomic Energy Commission insisted that it was premature to promote private investment in nuclear power. The more liberal policy was driven by Eisenhower's determination, eventually supported by AEC Chairman Lewis Strauss, to promote "the peaceful atom."[27]

The promotion of reactors for research and power abroad reached a peak in the summer of 1955, just as the plans for Euratom were being drawn up in Europe. In June, in a much-publicized commencement address at Pennsylvania State University (which had the first campus-based research reactor in the US), President Eisenhower characterized nuclear energy as a revolutionary new technology that "in its industrial and economic aspects, will likely bear no more resemblance to the age of steam than a jet-powered plane to an old-fashioned box kite." "Our Nation," he emphasized, "has no desire for a monopoly on the knowledge and practice of these possibilities." He offered friendly countries research reactors at half price, along with nuclear fuel to power them. The president was more reticent regarding power reactors, which involved classified data and were potentially an important market for American industry. Foreign scientists and engineers would be granted "access to and training in the technological processes of construction and operation for peaceful purposes" of nuclear power plants that their countries bought from the US to satisfy their energy needs.[28]

On July 18, 1955, Eisenhower attended a four-power summit conference at the United Nations' office in Geneva that was widely recognized as having temporarily defused tensions between the rival blocs. While in Geneva, he took time to visit the American exhibit at the forthcoming

International Conference on the Peaceful Uses of Atomic Energy, which had been designed to showcase the tangible benefits of his Atoms for Peace plan. A working research reactor of the type that would be available to friendly countries had been flown in from Oak Ridge and assembled in a chalet near the UN office in Geneva. Eisenhower took the opportunity to operate the reactor in the presence of about 150 representatives of the news media. The press could actually see the control rods being withdrawn to produce a chain reaction in the de-mineralized water that served as coolant, moderator, and radiation shield. Eisenhower told them that the US had already signed twenty-four bilateral agreements with foreign countries enabling them to acquire reactor technology.[29]

The International Conference on the Peaceful Uses of Atomic Energy began on August 8, 1955, ten years after atomic bombs were used on Japan.[30] Fourteen hundred delegates from seventy-three countries heard Homi J. Bhabha, Secretary to the Government of India's Department of Atomic Energy, enthusiastically proclaim: "For the full industrialization of the under-developed areas, for the continuation of our civilization and its further development, atomic energy is not merely an aid; it is an absolute necessity."[31] Three thousand scientific and technical papers, five hundred of them from the United States, were published in the proceedings of the conference, which was characterized by one British observer as "the most momentous scientific occasion the postwar world had ever seen."[32] Visitors to the American exhibit watched the reactor go critical before their very eyes. A series of panels illustrated some of the possibilities of the peaceful atom—notably the use of radioisotopes in agriculture, medicine, and industry. A trade fair held in Geneva at the same time showcased commercially available nuclear technology and emphasized the strength of Britain's nuclear industry. According to a Department of State official, there were "massive and to some extent competitive interchanges of information" thanks to the declassification of materials previously withheld by the McMahon Act.[33]

Kenneth Osgood has characterized Eisenhower's Atoms for Peace program, including the major effort to vigorously promote its benefits abroad, as a "carefully orchestrated psychological warfare campaign" that evolved into "quite possibly the largest single propaganda campaign ever conducted by the American government."[34] Eisenhower recruited Charles D. Jackson from the newsweekly *Time* to promote the campaign as a "direct challenge to Soviets' near monopoly of 'peace propaganda.'" As Jackson saw it, Atoms for Peace was just what the United States needed "to go on the moral ideological offensive against the Communists … and give it a bite and punch that would really register on both sides of the Iron Curtain."[35] However, the "rhetoric" surrounding the peaceful atom cannot simply be reduced to propaganda. Long-standing restrictive policies were revised. Significant changes to the law enabled the AEC to provide nuclear material, know-how, and training to foreign countries through bilateral agreements for research and power—about forty agreements to do so had been signed by 1961.[36] The door was opened for the commercialization of nuclear power by American industry. By February of 1956, about 25,000 technical reports had been reviewed, of which about a third were freely available and about a fourth were reclassified L (indicating limited clearance) and made available to American firms that wanted to enter the nuclear power field—as they did, gaining 90 percent of the reactor export market by the end of the 1960s.[37] To reduce Eisenhower's initiative to a struggle for hearts and minds, as Osgood suggests, is to overlook the significant measures the AEC took to disseminate nuclear materials and to share know-how throughout the world.

Atoms for Peace was a global "Marshall Plan" for atomic energy. The "hundreds of thousands of visitors" who flocked to see working models, colorful displays, educational panels, and nuclear equipment in one major city after the other—first in Geneva, then in Belgrade, Buenos Aires, Frankfurt, and Kyoto—were not gullible victims of American propaganda.[38] They were responding to the concrete opportunities made available by the US to propel them along the path to modernization that was promised by nuclear power, a path that proved to be far more rocky than anyone thought at the time. That same utopian dream, and the hope of fulfilling it with American help, inspired the founders of Euratom.

Relaunching Europe[39]

The movement toward European integration shuddered to a halt in August of 1954, when France's National Assembly refused to ratify the treaty establishing a European Defense Community. As an alternative to Germany's accession to NATO, this supranational body was supposed to create a pan-European military that would include Germany. The French opposed a system that denied them a national army under their control primarily to avoid arousing German resentment. For several months, any further talk of supranationality was political anathema.

Supranationality was soon back on the agenda, however, in part because of the prospect of establishing a European atomic energy "pool" with American help. In April of 1955, a senior official of the Coal and Steel Community told the Department of State that the peaceful use of the atom was the "most exciting and promising" platform for a new supranational initiative on the Continent, "since there were no vested interests in this field as yet and since it was a wave of the future which had captured people's imagination."[40]

In June of 1955, the Foreign Ministers of the Six met in Messina to see what measures could be taken to relaunch European integration. They established an Intergovernmental Committee (chaired by Belgium's Foreign Minister, Paul-Henri Spaak) to explore the possibility of achieving "horizontal" integration in a European common market and "vertical" integration in several functional sectors of the economy, including transport and atomic energy. Spaak was expected to report back later that year, and to present draft treaties and agreements for further consideration. He was assisted by Jean Monnet, who served as the president of the High Authority of the Coal and Steel Community from 1952 until he resigned in June of 1955. In October of 1955, Monnet formed an Action Committee to promote a United States of Europe. It sought both to build political alliances among the Six in favor of supranational projects and to advocate that agenda in Washington.

European integration was warmly supported by President Eisenhower and by Secretary of State Dulles as "a bold and practical step in the direction of a united Europe," something in which Eisenhower took a "keen personal interest."[41] Their general motives—peace and strength—ably complemented each other. Dulles told Germany's Minister of Foreign Affairs, Heinrich von Brentano, that "only troubles and wars could come from the continued division of Europe," and that "unification" or "integration" was imperative.[42] Eisenhower, at a meeting of the National Security Council (NSC) in November of 1955, spoke of "the great advantages of a more nearly united Europe—cultural, economic, moral, and otherwise." A "third great power bloc," the president told the NSC, "a solid power mass in Western Europe," would "ultimately attract to it all the Soviet satellites," removing a major threat to peace and allowing the US "to sit back and relax somewhat."[43] An integrated Europe, rather than squandering its potential in national rivalry, could devote its energies to building a major force on the front line of the Cold War, taking at least some of the burden of its defense off the shoulders of the United States.

The Department of State and the Atomic Energy Commission immediately disagreed on how best to respond to the Europeans' move toward creating an atomic "pool." Assistant Secretary of State for European Affairs Livingston Merchant favored a proactive approach. He wanted a common European authority, if one were to be established, to be treated as if it were a single state, so that it could qualify for support under the new, more generous terms of the 1954 Atomic Energy Act. He also suggested that Belgium be authorized to use its privileged relationship with the United States (it guaranteed the US a supply of uranium ore from its mines in the Congo) to attract other Continental powers into the pool.[44] Merchant also called on the AEC to restrict new bilateral agreements with any of the Six to research reactors, specifically refusing to enter into such arrangements for power reactors.[45] Through this combination of sharing and denial, Merchant hoped to make the supranational option more attractive to the Six.

Merchant's proposals were rejected at once by Gerard Smith, Dulles' Special Assistant for Atomic Energy Affairs. The AEC shared Smith's views. First, it was not at all clear that the 1954 Atomic Energy Act permitted cooperation in peaceful use with any multilateral organization other than the IAEA. Second, the bilateral nuclear power agreement that had just been signed with Belgium was "most important to our national security and to the defense of the Free World"[46] and could not be renegotiated without further careful consideration by both parties. Smith was emphatic that the United States should take no steps to support a European atomic energy authority. It was not yet clear what that venture involved or how, if at all, the United States could cooperate with it.

An interdepartmental group met a fortnight later at the Department of State to take stock of the situation and reached no definitive conclusions. It was thought unwise to try to "bend" Sections 123 and 124 of the 1954 Atomic Energy Act (which dealt with international cooperation) to allow for the transfer of fissionable material or classified information to a group of nations rather than to a single state or to the IAEA.[47] Even then, any agreement for cooperation in terms of Sections 123 and 124 would not be easy to implement. The Six, in view of their level of scientific and technological capability, would want "data which is on the frontier of research and development." Smith "could not imagine" the US sharing such data with them. The AEC, he argued, would cite the presence of many left-wing nuclear scientists in France, and the European scientific community would oppose the institution of invasive "security systems comparable to ours."[48]

For the next few months, the Department of State could do little but wait. Spaak's committee was having trouble completing its report. The AEC was reluctant to encourage closer ties with Europe.[49] In any case, Monnet and Spaak were "most emphatic" that the US "should continue to stay entirely in the background" as regards European integration, for fear that any public statement of support would be interpreted as a form of external pressure, if not blackmail.[50] No one had forgotten the outcry in Europe after Dulles' famous speech to the North Atlantic Council in December of 1953. Dulles had said then that failure to ratify the treaty creating the European Defense Community "would compel an agonizing reappraisal of United States policy" in the region. The next day he had suggested that the US would withdraw its security guarantees and leave Europe to "commit suicide" on its own if it did not support this supranational instrument for German re-armament.[51] The public facade of autonomy had to be maintained, notwithstanding European dependence on American nuclear know-how and materials.

Technological Leadership Emerges as a Political Weapon

> I think it would be most regrettable if, for whatever reason, atomic energy was not used by the US as a lever to obtain a further major step forward in the integration of Western Europe. As I see the problem, in its simplest terms, all the US has is a roughly three year head start over and above, of course, our tremendous overall engineering knowledge, personnel depth, etc. … I have long felt that it was within the power of the US to force almost any reasonable institutional form that it desired in this field, but only if it acted courageously and soon.
> —internal memo, Department of State, November 1955[52]

By the end of 1955, the Department of State was seriously considering a change of policy. Robert Barnett in the Office of European Regional Affairs circulated a memorandum on December 6 in which he emphasized that the United States could not afford "to remain comfortably mute for much longer." Concrete and dramatic action was needed, he asserted.[53] The president should be prepared to make an offer of technological support for a nuclear facility that the Europeans wanted, on condition that "the Community of the Six establishes by treaty an institution of sovereign authority and responsibility to administer the facility." To avoid being criticized again for meddling in European affairs, the president's integrative proposal should be conveyed in private to Paul-Henri Spaak (and, if it seemed necessary, to other leaders) in advance of any public statement. The Belgian Foreign Minister and his colleagues would be encouraged to "formulate a concrete proposal to the United States Government to which this [i.e., Eisenhower's offer] would, in effect, be a response."[54]

Howard Robinson at the US embassy in Paris added his voice to those advocating the use of atomic energy as a technological bargaining chip. In December of 1955, writing to Assistant Secretary of State for Policy Planning Robert Bowie, he asserted that "the march of events since the atomic meeting in Geneva was forcing the US into making basic atomic policy decisions which may be drastically different from those under which we have hitherto been operating." When Dean Acheson was Secretary of State, Robinson pointed out, "it was departmental working policy that atomic energy coin should only be paid out in cases of atomic energy benefit." That policy, Robinson argued, should be re-evaluated. As Robinson put it, "the US possesses at this moment a political weapon in atomic energy, the value of which [was] currently at its maximum." It should be willing to use this leadership to "extract non-atomic political benefits" in return for sharing atomic power with Western Europe. And it should move fast, for although it was "too soon to prophesy how quickly this coin will deteriorate in value" it was "quite certain" that basic decisions to use it as an instrument to "foster or to force non-atomic political results" would "surely have to be taken very quickly."[55]

Robinson's argument is echoed in the quotation at the head of this section. Why were "drastic" changes in basic US policy needed "quickly"? Why did the Department of State need to act "courageously and soon"? One reason, as we have just seen, was that the declassification of large amounts of nuclear information that had been stimulated by the August 1955 conference in Geneva threatened to debase the value of the "atomic coin": the nuclear knowledge gap between the US and other industrial powers was shrinking rapidly. A second reason was that the political situation in Europe was in a state of flux. As regards Euratom, two incompatible models of nuclear cooperation were battling for supremacy. One was tightly integrative, in line with the Messina framework. The other was multilateral in the framework of the OEEC—and strong forces in Britain and Germany were lining up in favor of it. By the end of 1955, neither American officials nor Spaak's committee had come forward with a clear choice between the two models. Hence the urgency in Washington. The movement toward European "integration" was at a turning point. Direct action was needed. The United States' "three-year head start" in atomic energy still provided it with a "lever" that the Department of State could use "to force almost any reasonable institutional form that it desired in this field." The techno-political instruments were at hand to impede European governments from veering off in a multilateral direction that would be at variance with the foreign-policy preferences of Eisenhower and Dulles, perhaps irreversibly scuttling their dream of a United States of Europe.

Competing Models, Conflicting Priorities, and British "Sabotage"

Several approaches to managing civilian nuclear matters in Europe were explored in the closing months of 1955 and were made public early in 1956.[56] The OEEC Council took the lead. It invited Louis Armand, director general of the French national rail system and director of industrial applications in the Commissariat à l'Énergie Atomique, to report on the possibilities of European nuclear collaboration for energy production within a multilateral framework.[57] Armand disappointed the Council by gravitating toward an integrative approach, and a new report was quickly published in January of 1956 (the Nicolaidis report), even before official consideration and approval.[58] It favored a loose and voluntary arrangement to facilitate little more than technical cooperation in the field of nuclear energy. There would be no common budget. Each country would secure fissionable materials through bilateral agreements with other countries. Countries would collaborate selectively with partners of their choice in activities of mutual interest. Common installations could be developed (for uranium enrichment, for example), but they would be managed by independent companies. Indeed, centralized control would be exercised only to keep fissionable materials produced in such plants from being diverted to weapons programs.

A different approach was suggested by the study group organized under the aegis of Spaak's Intergovernmental Committee. It met for the last time in November of 1955, and it announced its plans publicly in February of 1956. The study group's approach was the antithesis of the OEEC's scheme, though it was still vague about the degree of integration it favored. Although each member state could pursue its own national program, a central body would acquire and control all nuclear material throughout the fuel cycle; there would be no bilateral government-to-government agreements. Installations, such as isotope-separation plants, would be jointly owned and financed from a common budget. A common market for nuclear material and equipment would be established. The Messina group, like the OEEC, was silent on the issue of nuclear weapons, though it did suggest that all nuclear activities should be for peaceful use only, at least in the immediate future.

The Messina group stopped short of proposing a specific institutional framework for the new organization. Monnet's Action Committee grasped that nettle on January 18, 1956, when it unanimously approved a draft resolution calling for the creation of supranational European Atomic Energy Community along with a European Atomic Energy Commission modeled on the American AEC. For Monnet, Euratom's prime objective was to promote the peaceful use of nuclear energy in Europe through a pooled supranational effort intended to keep participants from developing national nuclear programs. The organization would jointly own all fissile nuclear material: no country could acquire fuel through an independent bilateral agreement. It would control the entire nuclear fuel cycle through joint ownership of enrichment plants upstream from the power reactors, as well as through joint ownership of chemical separation plants for processing used fuel downstream from power production.[59] All members would have to accept a system of inspection and control that would ensure that material produced in national programs was not being diverted to military programs. Observers from Britain, the United States, and the Soviet Union would have the right to confirm compliance.[60] Finally, all parties to the Euratom treaty were expected to renounce the right to make nuclear weapons. The new socialist Prime Minister of France, Guy Mollet, who worked closely with Jean Monnet, publicly supported this radical proposal, which was bound to raise hackles in Paris. "We will ask that the member states of Euratom take a solemn engagement to renounce the use of the atom for military purposes," Mollet wrote soon after taking office in February of 1956.[61]

One particularly important argument that favored Monnet's supranational model was the claim that tight political integration would enhance security by building a high wall around the diversion of materials from peaceful to military objectives. Paul-Henri Spaak visited London in November of 1955 and made the case with admirable lucidity. According to one account,

> He was very much impressed by the need to find some means of controlling the supply and use of fissile material in the interests of security. If the need for control was accepted, the need for a controller followed. It was not easy to see how a satisfactory form of control could be evolved without some measure of supranationality. From this angle an organization modeled on the concept of Euratom which might have monopoly powers in relation to the purchase and supply and distribution of raw materials, and which might operate in common certain key processing plants, could be an effective instrument for accounting for and regulating the use of fissile material.[62]

By contrast, the far looser arrangements foreseen by the OEEC, Spaak insisted, would increase the proliferation risk, since material flows would be far more difficult to regulate. The European project for political integration around an atomic pool was also a powerful instrument for curbing proliferation.

The British were not persuaded. The initiatives taken in Messina as regards both economic and atomic integration led to extensive discussion inside Britain over how best to respond. The consensus that rapidly emerged was that the UK had no interest in joining an integrated European economic system with a common external trade policy that required a measure of central coordination. The UK preferred instead to promote a free-trade area in which individual partners would negotiate external tariffs themselves. They also agreed not to join a European atomic energy organization. However—and here was the rub—whereas previously the British attitude toward European integration had been one of benevolent non-commitment, now the British government was not content to ignore the push toward integration on the Continent. In October of 1955, at a meeting of the Mutual Aid Committee (a group of officials who reported to the Cabinet), it was resolved that the common market "if possible, should be frustrated," since "we cannot count on the project collapsing of its own accord." The Six should be attracted into the orbit of the OEEC, where their integrating ambitions would be quickly snuffed out. The strategy (as Gladwyn Jebb, Britain's ambassador to Paris put it) was to "embrace destructively."[63]

On November 17, 1955, the British position was officially conveyed to Washington, where it generated the consternation described above. It was simultaneously conveyed to Bonn, capital of the FRG. Britain's preference for a free-trade area was shared by Germany's Economics Minister, Ludwig Erhard, by its newly appointed Federal Minister for Atomic Affairs, Franz Josef Strauss, and by German industrialists who were opposed to imposing "socialistic" and "protectionist" practices on the Continent. It was hoped that they might persuade Chancellor Konrad Adenauer to tilt in favor of the looser OEEC arrangement. As the Foreign Office put it at the time, "If we were to give them a lead, the Germans might … adopt a more realistic policy … and decide not to join a common market and concentrate on co-operation through the OEEC."[64] There were other advantages too, as one Treasury official noted: "If we do not want to see a common market set up, and if we want to avoid being blamed for its failure, it would suit us well to persuade the Germans to lean towards OEEC."[65]

The decision by the British to actively oppose European integration, and to try to split the Six by wooing Germany into their camp, was a dramatic reversal of policy. As Piers Ludlow puts it, "in 1955, unlike five years earlier, London no longer felt strong enough to watch benignly from the margins as six of their Continental neighbours went ahead with economic integration." Instead, the British sought, briefly and abortively, to "strangle the EEC at birth" by suggesting that European governments had to choose between integration with the Six without Britain and multilateral cooperation with Britain in the OEEC.[66]

The Roots of British Opposition to Euratom, and Dulles' Response

Britain's hostility to supranational arrangements in Western Europe is well known and doesn't require detailed elaboration here. Integration, it was thought, would unduly privilege one of the three circles of British postwar engagement and influence (Continental Europe) at the expense of the other two (the Commonwealth and the United States). The dilution of sovereignty that tight integration entailed was incompatible with Britain's perception of itself as a major power that could significantly influence the course of world events. A common market, in particular, threatened to erect high external tariffs against British imports, and violated the hallowed principles of free trade. Britain's reasons for resisting integration into Euratom resonated with some of these concerns, but assumed forms specific to nuclear energy.

First, the UK Atomic Energy Authority (UKAEA) could not accept the requirement that Euratom control the supply of uranium ore and fissionable material to its member states, as the American AEC did.[67] Britain, along with the US and Belgium, had privileged access to uranium through the Combined Development Trust (CDT).[68] The British Joint Chiefs of Staff were making demands for an expanded and diverse weapons program that could not be satisfied. The needs of British industry also had to be considered. Britain wanted to maintain control over its nuclear fuel, a scarce commodity that was much in demand by the military and an important bargaining tool when negotiating reactor sales.

A more fundamental obstacle was the dual-use character of the British nuclear power program. From the outset of that program, the UK had sought to build reactors that would generate useful—and eventually commercially viable—nuclear power while also producing plutonium for nuclear weapons. The new nuclear power station at Calder Hall, opened by the Queen on October 17, 1956, had been "built as a requirement for more military plutonium and as an experiment to investigate the possibilities of adapting nuclear energy to the production of electrical power quickly, cheaply and efficiently."[69] Officials repeatedly pointed out that the dual use of the country's reactors for breeding plutonium and for electricity production meant that in Britain the civilian program was "closely integrated" with the military program.[70] As Sir Harold Caccia in the Foreign Office put it, "the fact that our civil programme is carried on the back of our military programme effectively rules out any question of our becoming a member of Euratom." "We could not, in any case, put such a vital part of our military effort under international control," Caccia emphasized.[71]

Finally, as was noted by American reporters at a trade fair that accompanied the 1955 Atoms for Peace Conference in Geneva, Britain had a "vigorous lead" over the United States and over its Continental neighbors in the commercial exploitation of nuclear power.[72] The UKAEA was determined that no institutional arrangements should rob it of that position and the access to markets that it allowed. In ten years Britain had spent about £200 million on acquiring a technical capability in the field of nuclear science and engineering.[73] As a result, as William Strath, a senior civil servant in the Treasury involved in the making of nuclear policy noted, Britain had "a technological lead over most of the European countries who [would] be our competitors in the future." She had to "cash in" on her position "for as long it [was] possible to do so."[74] Any form of international collaboration carried risks that Britain was loath to take and made demands for sharing knowledge and materials that she could not satisfy. An integrative arrangement was particularly disadvantageous. Britain's leadership would be squandered in a pooled program in which "no individual country can hold back from co-operation on certain activities where it feels it has a specially strong position, and confine its co-operation to those developments where it feels it has more to gain than to lose," as Strath put it to the chairman of the UKAEA, Sir Edwin Plowden.[75] A supranational organization, then, would oblige Britain to give away its scientific, technological, and commercial leadership to partners who had little if anything to give in return. By contrast, the Organization for European Economic Cooperation was "a flexible system of multinational cooperation which would embrace different countries at different stages of [technological] development."[76] It was dominated by economic considerations, it respected the independence of sovereign states, and it allowed each partner to share peaceful nuclear knowledge and materials selectively with those of its own choosing and in line with its own interests.

Dulles could not ignore these developments, which were contrary to his convictions and to those of President Eisenhower. On December 10, 1955, the Secretary of State wrote a personal letter to the British Foreign Secretary, Harold Macmillan, regarding the most desirable policies for promoting the peaceful use of nuclear energy in Europe.[77] He escalated the issue to one of geopolitical significance. Western "unity and strength" were crucial in the face of subversive Soviet attempts to "divide us and, through so-called 'peaceful competition,' to step up the battle for men's minds." Germany had to be tied into "new forms of integration" in the fields of nuclear and conventional energy that would not be at the expense of continued cooperation through the OEEC, but rather would be compatible with it. Dulles hoped that he "could count on" British support for this approach. There might be differences of emphases between the US and the UK, he admitted, but Washington and London should not give "conflicting advice to the Continental Europeans with respect to the movement towards closer unity."

The British were somewhat shocked by the hostile response to their preference for the OEEC. No one expected them to join a supranational body; they were simply being asked not to impede it. Instead, as Dulles reported to Eisenhower in mid December, the British were "working hard to block the development of the European idea in terms of atomic energy and a common market."[78] At a meeting of the Council of the Western European Union (WEU) on December 14, 1955, Paul-Henri Spaak and the Dutch Foreign Minister, Jan Beyen, strongly criticized Harold Macmillan for making the "tragic blunder" of suggesting that one had to choose between the OEEC and a common market, rather than recognizing that each had a role to play. They feared that British approaches to Germany and to France might well "spell the end of all hope of progress towards European integration."[79] Dulles repeated his objection to Macmillan at the NATO ministerial session of the North Atlantic Council in Paris the next day. Chastened, the British seemed to draw back. At the end of the year all their embassies in Western Europe were instructed to "stop plotting" against the Six, and to show reserved diplomatic support instead.[80] Britain could not afford to be at loggerheads with the United States on a matter in which both Eisenhower and Dulles were so heavily invested.

The retreat was mere window dressing, however. In January of 1956, Franz Jozef Strauss visited London to discuss Germany's options for developing nuclear energy. Several weeks earlier he had visited the American ambassador to the FRG, James Conant, a chemist who had been president of Harvard University in the 1930s. Strauss had expressed his strong objection to the centralized control over the fuel cycle and of German industry that was foreseen for Euratom.[81] The far less restrictive approach being advocated by Britain obviously appealed to him. He was also impressed by the focus of the UK's civilian program that seemed better suited to the FRG's immediate situation.[82] The United States, with ample cheap conventional power, had developed nuclear reactors for submarine propulsion and had paid little attention to their commercial use. But Britain, like Germany, lacked cheap coal, and had devised a nuclear power program to satisfy the need to generate more electricity. Britain would soon be producing electricity at Calder Hall, while the US was still engaged in "long-term industrial experiments." A month later the British ambassador to the FRG told Strauss that the UK would be pleased to help Germany develop its national program with bilateral agreements—the very measure that Spaak and Monnet feared would sound the death knell of Euratom, and that the US Department of State was desperately trying to obstruct.[83] In short, Britain was more than willing to collaborate with Germany in the framework of the OEEC, thereby forging tight bonds between the two civilian nuclear power programs and seriously undermining the push toward integration by the Six.

Dulles Goes on the Offensive

Dulles returned to Washington in December of 1955 determined to do something to change the terms of the debate in Europe. He suggested, and President Eisenhower immediately agreed, that the Department of State and the Atomic Energy Commission urgently study what moves the US could make in the atomic energy field "to encourage six-country integration."[84] The "German problem" dominated Dulles' thinking. He doubted that Chancellor Konrad Adenauer would abide by his 1954 commitment not to produce nuclear weapons on German soil. And what would happen when Adenauer was gone? Would the Soviet Union use its control over East Germany to trade German reunification for German "neutrality"? Dulles emphasized this point when he wrote to Macmillan in December. There was, the Secretary of State wrote, "the danger that the appeal of reunification will, over a period of time, become so strong in Germany as to give rise to the temptation to discard the associations with the West in an effort to advance reunification which would at best result in a neutral Germany and at worst in an eastern-oriented Germany."[85] The most durable way to "prevent this possibility arising," as Dulles put it to Macmillan, was "to tie Germany into the whole complex of Western institutions—military, political and economic—and to so command her loyalties, that neutrality or orientation to the East will be commonly accepted as unthinkable."[86] That was a tall order, Dulles conceded, but he saw no alternative. Only the quest for European integration was exciting enough to "occupy their [German] minds in the inevitable period of waiting that lies ahead on the reunification question."[87]

Germany also had to be a member state of an integrated Euratom. The nuclear platform was spearheading European unity. And it was an invaluable instrument of non-proliferation. As Dulles put it to Eisenhower early in January 1956, if the Six established, "an integrated institution possessing effective central and inspection authority in the field of peaceful uses of atomic energy, control over military uses of atomic energy by these six countries would be simplified … ." The value of Euratom as a collective responsible for safeguarding nuclear materials, and ensuring that they were not diverted clandestinely to weapons programs, was a second major reason why a supranational solution was imperative.

On January 25, 1956, Secretary Dulles and other senior members of the Department of State met with representatives of the Atomic Energy Commission, including its chairman, Lewis Strauss. The aim of the meeting was to acquire "authoritative technical advice as to what the US *could* do," by sharing knowledge and technology, to "catalyze development [of] effective supranational integration in the atomic energy field." Existing legal, economic, or other obstacles were to be ignored for the moment: they could be dealt with later.[88] What technical support could the AEC offer to an integrated body before the "wasting asset" of America's lead in civilian nuclear matters lost its bargaining power?[89]

Lewis Strauss drew back. Fearing security leaks, he was reluctant to share classified information with the French—and hence with Euratom. He was also bothered by the risks of proliferation. A suggestion that the US help the Europeans build a gas-diffusion plant to enrich uranium was rejected outright. Disappointed, the Department of State urged AEC officials to put aside their doubts. The Europeans would go ahead with "increasingly active and successful programs in the field of atomic energy" anyway, and the US would have to move fast to influence their program in desirable ways. "Admiral Strauss said that he wanted to cooperate with the Department to the hilt—up to the legal limit. The Secretary said that he hoped that the Commission might exceed those limits and consider all possibilities … ." Indeed, a report of what could be done was needed quickly "since any changes in the law should be submitted to the Congress well before its adjournment … ." The Department of State wanted the AEC to think creatively and was even willing to ask Congress to expand the scope of collaboration beyond what was currently allowed by law to provide significant incentives to European governments if they established Euratom.[90]

Reporting back to officials in the Department of State in April, Lewis Strauss spelled out the "materials, information, services and special privileges" that the AEC could offer as an "incentive in the creation of an integrated European atomic energy industry." The offer included sharing uranium enriched up to 20 percent in the fissile U^{235} isotope, as well as plutonium and U^{233}. The AEC would share all technology up to the level of "confidential" for various plants required in the power reactor complex (fuel fabrication, chemical separation, and so on). It could offer assistance in education and training. And it would secure the help of American industry in establishing "cooperative, joint development programs as mutually agreed." Strauss went as far as to suggest that the Department of State, in promoting this agenda, propose that the "Community of Six" borrow funds from the Export-Import Bank, the International Bank for Reconstruction and Development, and other financial institutions. All of this was conditional on granting "US access to the country ... for the purpose of observing compliance with the assurances of no diversion for military purposes."[91]

Thus armed, Dulles was now in a position to act. The Europeans still weren't able to agree on the basic outlines of Euratom. Perhaps fearing that this might be his last chance to influence the course of events, Dulles decided to take a bold stand in anticipation of the next meeting of the Foreign Ministers, which was to be held on May 29 and 30 in Venice. Five days before the meeting he circulated a major policy statement to American embassies in the Six and in London. Drafted by Robert Schaetzel and Harland Cleveland in the Department of State, it unambiguously spelled out the American position.[92]

The US, Dulles' telegram read, supported European efforts to "create integrated nuclear community." It was a "possible decisive contribution" to more "general integration" that would help to tie Germany "organically to the West" and to strengthen the Atlantic community. It would "submerge" Franco-German rivalry by creating "intimate common interest" in the field of nuclear development.[93] If the "integrated nuclear community" had adequate control powers, it would be the best way to build a system of safeguards against diversion in the one area of the world in which nuclear development was likely to occur in the near future. And by "merging the scientific and industrial potential of [the] Six" it seemed to offer the best chance for the development of a nuclear industry in Continental Europe. Reiterating earlier decisions, Dulles then indicated how the US planned to use its technological advantages in the nuclear domain as a political weapon to achieve its objectives. The US would do so by giving preferential treatment to a Euratom that satisfied its criteria for integration—that is, "the US could make available substantially greater resources and adopt attitude of substantially greater liberality toward a real integrated community possessing effective common authority and responsibility than would be possible for countries separately." This condition made it imperative that Euratom "have authority over [nuclear] fuel which if not ownership, is as complete as if Euratom owned fuel." Embassies were also informed that the US opposed German industrialists' demands for the right to purchase and own fuel, for the right to arrange bilateral agreements with third parties, and for linking the establishment of Euratom explicitly with that of a Common Market. Host governments should be told that if the draft treaty for Euratom did not meet these requirements it might "raise problems with respect future ability US to cooperate substantially Euratom." The die was cast: there was a price to pay for American scientific and technological collaboration in this domain. If the Europeans wanted the US to share significant nuclear knowledge and materials with them they would have to rally behind an integrated structure with "common responsibility and authority" for a joint civilian nuclear energy program.

Dulles' offer had little immediate effect, except to anger the German Minister of Atomic Affairs even more. Indeed, the divisions between Germany and France on the best institutional shape for Euratom were still so deep that Dulles' gestures could barely dent them. In fact, as we shall see in the next two sections, the pressures in Bonn and Paris to stay clear of Euratom were still so significant that it took a shock from without to impress the need for European integration on the minds of German and French leaders. Then, and only then, could they look seriously at the technological incentives the Atomic Energy Commission and the Department of State were offering them in return for supranationality.

The Sources of German Opposition

If Germany is to support a real supranational authority, a very large carrot in the form of what the US is offering would have to be put out in front of the mouths of the six donkeys.
—James Conant to Livingston Merchant, February 1956[94]

Minister of Atomic Affairs Franz Josef Strauss had important meetings with US Ambassador James Conant in Bonn in February of 1956, and with Dulles and the Department of State in May, shortly before the Venice meeting. He arrived in Washington, determined that German industry compete with France and Britain in the field of atomic energy within six or seven years, and with a draft law on atomic energy in his pocket. With a view to producing as much as 25 billion kilowatt-hours of electricity from atomic energy by 1970, Strauss hoped to place three orders for research reactors with American firms.[95]

Strauss came to Washington in May to impress his objections to a Monnet-style Euratom on the American authorities. Germany preferred a scheme that "combined" Euratom and the OEEC in an arrangement that included Britain. Without the UK, Strauss said, Euratom would "consist of five blinds and one half-blind"—the latter being France, which had a vibrant nuclear program.[96] Strauss was also emphatic that no scheme for European integration could proceed piecemeal: agreement on Euratom alone without steps being taken toward a common market were unacceptable, their treaties should be ratified together, and if they were decoupled "he would immediately leave the Cabinet," the majority of Adenauer's ministers would join him, and the government of the FRG would collapse.[97]

Strauss had two strong objections to Jean Monnet's vision for a supranational Euratom. First, and fundamentally, he saw it as a French-led socialist maneuver intended to keep German industry in check.[98] Monnet's original proposal—"agency possession and monopoly of the purchase of nuclear fuel, agency power to permit or veto plant designs, and the elimination of bilateral treaties between individual member states and non-members"—was "absolutely unacceptable to the German Cabinet and parliament," he told the Department of State in May.[99] Strauss wanted private industry to own fissionable materials and to develop nuclear power with minimal government intrusion and with even less intrusion by an international bureaucracy that might impose tight regulations on its behavior.[100] Strauss was also deeply disturbed by the provisions being made to reveal sensitive technical knowledge to all the partners in Euratom. This "would be tantamount to handing over to the Soviet Union on a silver platter all of the products of Euratom science in the atomic field," he insisted. The bureaucracies in France and Italy, he claimed, "were riddled with Communists, and there would be absolutely no control over classified information on atomic questions."[101]

Second, Strauss disliked Monnet's suggestion that Euratom's member states were obliged to renounce nuclear weapons. He was skeptical about France's generally positive attitude to Euratom so soon after that country had killed proposals for the European Defense Community, and was convinced that France's aim was "really to ensure that Germany did not embark on a military program."[102] He resented the implication that Adenauer's 1954 assurances that the FRG would not produce nuclear weapons on its soil were not to be trusted.[103] By denying the Six the right to produce nuclear weapons, Monnet and his socialist allies were taking an "important step towards neutralization." Europe would become a "second-class military vassal" of the United States and Britain, dependent on them for the essential weapons of modern warfare. The Soviets, for their part "would be delighted to hear that Continental Europe had given up all chance of being a real actor in world politics by voluntarily confining itself to military forces armed with broomsticks and machine guns in an age of modern weapons."[104] Strauss did not openly favor a German nuclear-weapons program, but he was emphatic that the Continent should be free to develop its own nuclear capability.

These objections did not unduly perturb Dulles and the Department of State. The accusations that Euratom's centralized control over nuclear materials violated the principles of free enterprise fell on deaf ears. As one of the American authors of the 1946 McMahon Act explained, that law established "in the midst of our privately controlled economy a socialist island with undefined and possibly expanding frontiers." He was referring to sections of the act that gave the AEC "absolute control" over all fissionable materials, over all facilities that used or produced such material, and over all information related to nuclear technology.[105] In the US the needs of the national-security state trumped the principles of laissez-faire, and it was only with great difficulty that the Eisenhower administration gradually relaxed some constraints to enable private industry to exploit "the peaceful atom" for power. Even then it retained tight control over nuclear materials throughout the fuel cycle. Euratom need not have "socialistic results," one official in the Department of State told Minister Strauss; the US had "a strong Commission with strong controls," and it was "developing a private program."[106] German industrialists' claims that Euratom was a socialist plot to control their entrepreneurial initiative, and their demands for a far laxer regime, elicited little sympathy in Washington.

The question of weapons was more complex. No one quite trusted Germany, notwithstanding Adenauer's 1954 assurances.[107] Ambassador Conant warned Livingston Merchant that German nationalists, including the physicist Werner Heisenberg, might renege on Adenauer's commitment. He claimed that Heisenberg was prey to an inferiority complex resulting from his failure to develop a bomb for Germany during World War II.[108] Conant also remarked that, although "all Germans preface their remarks about atomic energy by disclaiming any intention to manufacture weapons," he detected "signs of uncertainty" when he spoke to them.[109] Eisenhower's New Look policy considered nuclear weapons to be as available for use as any other munitions, and to be deployable from bases in Western Europe in the event of hostilities.[110] In Conant's view, the FRG would not tolerate being kept in a position of inferiority, indefinitely forbidden to produce nuclear weapons of its own, even as US policy demanded that tactical nuclear weapons supplement conventional ground forces in Europe. When Dulles met with Strauss in May of 1956 he insisted over the latter's objections that the control system in Euratom was needed to ensure that nuclear material would not be diverted from peaceful to military purposes. A "multiplicity of uncontrolled national atomic developments" was totally unacceptable, he told Strauss. Adenauer and Brentano agreed with him, at least in public.[111]

It must be emphasized that Dulles did not object to national atomic weapons programs as such. It was their "uncontrolled" development that bothered him. Indeed, Dulles had already conceded that France would have to be allowed to develop a nuclear weapon if it were to join Euratom. The French parliament would never ratify the treaty otherwise. This point was specifically raised with Dulles at a dinner with senior French officials in February of 1956.[112] His commitment to integration overrode his concerns about proliferation. "I suggested," he wrote immediately afterwards, "that in connection with the UN Atomic Energy Agency there might be an agreement that 'fourth countries' would not make atomic weapons for a period of time—say five years—during which an effort would be made to eliminate these weapons by agreement between the United States, the Soviet Union and the United Kingdom." As Trachtenberg emphasizes, Dulles was not against Europe's achieving "atomic independence" so long as it was within an integrated framework. Eisenhower was inclined to be even more liberal and allow Europeans effective control over nuclear weapons on their soil, notwithstanding strong opposition in the AEC.[113] Euratom's credentials as an instrument for non-proliferation were thus dented within nine months of the *relance* at Messina.

Franz Josef Strauss and his colleagues were aware of the broad outlines of Dulles' support for a supranational Euratom that he officially circulated in May of 1956 to embassies in the Six. Strauss was incensed by Dulles's initiative. According to a member of the American delegation who chatted with him over lunch, Strauss accused the US of making exactly the same mistake it had made with the European Defense Community: "seeking a political objective through the application of pressure and influence." To make integration a condition for meaningful collaboration would earn the US "nothing but ingratitude in Europe," Strauss went on, adding that such policies "have lost you North Africa and the Middle East, and will probably lose you all the uncommitted countries of Asia."[114] Whether or not Dulles heard this, he did not hesitate to give Strauss a drubbing. If one wanted to understand why the American president supported steps toward a "unified Europe," he told Strauss, one simply needed to remember that "the divisions in Western Europe have brought Western society close to extinction." The US may not have been concerned before, but now it was deeply involved. "We had lost hundreds and thousands of men and hundreds of billions of dollars as a result of wars starting in Europe," Dulles told Strauss. "We cannot sit aside and say nothing," he continued, "while the structure which has been the world's greatest war hazard is being rebuilt."[115] Dulles would not yield: Germany would have to join a supranational Euratom.

France's Anxieties about Euratom

In 1955, France was far ahead of any other Continental country in nuclear investment and planning. It had established a Commissariat à l'Énergie Atomique (CEA) immediately after World War II, drawing on its strong tradition in nuclear physics and on the skills of its scientists who had worked in Canada on the Manhattan Project. The government's five-year plan drawn up in 1952—the year in which Britain successfully tested its first atomic bomb—already raised the possibility of developing atomic weapons. By 1954, the nuclear-weapons option had become more definite in response to Eisenhower's New Look policy. According to Georges-Henri Soutou, the French reasoned that if nuclear weapons were going to be deployed at the onset of hostilities in Europe, and not only after conventional forces had been routed, then they could only hope to influence their own military destiny if they too had a nuclear arsenal.[116] However, when Prime Minister Pierre Mendès France thought to embark on a major bomb program in December of 1954, he was stopped by Francis Perrin, the Scientific Director of the CEA. Perrin pointed out to him that France still lacked the scientific, technological, and industrial infrastructure for an independent weapons program. Anyway, 600 employees of the CEA had already signed a petition refusing to work directly on weapons of mass destruction.[117] Perrin insisted that the "common trunk" shared by civilian nuclear power and nuclear armaments would have to be developed before France could build its own weapons. Hence the idea arose that it would be in France's interest to pool resources in the civilian domain with European partners so as to advance more rapidly in the development of nuclear power, *thereby* also releasing scarce personnel and funds for an expensive military program.

In view of these priorities, it is not surprising that for much of 1955 the French favored working in the framework of the OEEC to build their "common trunk." It would ensure their freedom of action, it posed no threat to their weapons program, and it gave them an opportunity to collaborate with the British, from whom they had much to gain. This point of view was gradually supplanted during the course of 1955 by support for a European atomic pool of the Six. There were two main reasons for that change.[118]

First, it was thought that the Euratom option would enable France to have access to uranium ore in the Belgian Congo and to American technical know-how. In June of 1955, immediately after the Messina meeting, French officials told C. Douglas Dillon, the American ambassador in Paris, that the pending agreement between Belgium and the US offering special privileges to Brussels in the field of civilian nuclear power would endanger a European pool unless those privileges were extended to all participants.[119] The French hoped that Belgium's advantageous position in the Combined Development Trust, which controlled the distribution of about 97 percent of the known deposits of uranium ore in the world (notably those in Katanga), would accrue to the Six collectively.[120] So too should the classified know-how on nuclear power reactors available to Belgium in its bilateral agreement with the AEC that was signed on June 15, 1955. And indeed Foreign Minister Spaak, a convinced integrationist, proved more than willing to use these Belgian assets, with American agreement, as an incentive to others to join in a supranational Euratom.

The second French concern was that in an OEEC-like arrangement Germany could strike favorable bilateral deals with the US Atomic Energy Commission that would enable it to rapidly overtake France in the industrial production of nuclear power. By contrast, if Germany were obliged to pool its resources in Euratom, its autonomy would be constrained, its strengths would be shared with its partners, and a close eye could be kept on the development of its nuclear program. In short, the French hoped, by pooling capital, scientists, engineers, and industrial capability through a Euratom of the Six, to take a commanding position in civilian nuclear matters on the Continent with the help of, but also at the expense of, Germany. In so doing, they would free up scarce resources for their nuclear-weapons program and secure a degree of strategic independence from the US that would put them on the same footing as the UK.[121]

Of course, Monnet's conception of Euratom, initially supported by the Socialist Prime Minister Guy Mollet (who took office early in 1956), required that member states renounce the development of nuclear weapons. That requirement was doomed to fail. Dulles' solution to the problem—that "fourth countries" *temporarily* desist (for, say, five years) from actually increasing the world's nuclear stockpile—was taken up by Spaak and welcomed by most of the Foreign Ministers who met in Venice in May. They realized that an arrangement of this kind was needed to keep the French National Assembly from again scuttling a move toward European integration. Indeed, in June of 1956, seeking to win assent for continuing negotiations on Euratom, Mollet reluctantly invoked and even relaxed that condition. France would not test a bomb for four years, and after that it was free to proceed as if Euratom did not exist.[122]

By August of 1956, a mood of gloom had descended on the Department of State. The situation in Euratom was discouraging. The offer of support that Dulles made just before the meeting of the Six in Venice in May appeared to have made no difference. France demanded the right for all states, except Germany, to continue research and development on weapons during the four-year moratorium before a bomb could actually be tested (just the time the CEA calculated it would need anyway). German industrialists remained implacably hostile to Euratom's exercising common ownership over fissionable material, and Adenauer seemed to be leaning their way. The French and German governments were both drawing back from the supranational model, arguing that participating states and companies should be free to procure nuclear materials outside Euratom—in fact, France suggested that Euratom should control no more than 20 percent of its atomic energy program. And then there was the question of linkage: Germany (with some support from Belgium and the Netherlands) continued to insist, and France to refuse, that the projects for Euratom and the Common Market proceed in unison or not at all.[123]

Bob Schaetzel summed up the situation as he saw it. The compromises being struck by the Six all worked "in the direction of creating an institution which promises slight encouragement if any to integration, more of a cooperative than a community scheme, equivocal to the point of disaster on the weapons question, and with the resistance to outside supervision of the control mechanism which raises substantial doubts as to its effectiveness in this important area." "The drift towards cooperation rather than integration," Schaetzel went on, "the emphasis on separate national programs as opposed to a strong community approach," all these developments pointed toward the emergence of a "non-action enterprise; an enterprise whose principal function would be intrinsically negative responsibility for safeguard and control." Euratom was being evacuated of its supranational, non-proliferation characteristics, and there was not much that the US could do about it. "We have been constantly rebuffed" by the proponents of Euratom, wrote Schaetzel, for they feared that "any direct involvement of the US in the negotiating process would be highly damaging."[124] Schaetzel did not know that the entire situation was soon to change dramatically, with help coming from an unexpected quarter.

2

The United States and Euratom, 1957–58: Constructing a Joint Program for Nuclear Power

Bob Schaetzel's gloomy prognosis in August of 1956 was dispelled by two unexpected events: Egypt's nationalization of the Suez Canal earlier that year and the USSR's launch of the earth satellite Sputnik 1 in October of 1957. The former accelerated the intergovernmental negotiations that led to the signing of the Treaties of Rome in March of 1957. The latter facilitated a favorable opinion by the Joint Committee on Atomic Energy (JCAE) and Congress on a new joint US-Euratom program that was ratified in August of 1958. This program combined the subsidized export of American reactors in Europe with a co-funded R&D exercise intended to push the state of the art in nuclear power technology. It was launched in the face of violent opposition from the British government (on commercial grounds) and from the International Atomic Energy Agency (which objected to Euratom's having the right to implement its own safeguards system—i.e., to "self-inspection"). Eisenhower and Dulles held firm: promoting European integration around a technological platform inspired by the Atoms for Peace program trumped all objections. Their firm convictions enabled them to take a pragmatic approach to Euratom's retreat from full supranationality and to France's parallel development of nuclear weapons. The JCAE was similarly indifferent, some members even empathizing with France's military ambitions. American soft power, already a wasting asset in the atomic energy field by 1957, was no match for national fervor in a field so charged with political and military valences.

The Breakthrough: The Suez Debacle[1]

In July of 1956, President Gamal Abdel Nasser of Egypt nationalized the Suez Canal, through which about 70 percent of Western Europe's oil had to pass. Britain had a 44 percent stake in the Suez Canal Company and saw it as one of a number of strategic assets that secured its influence in the Mediterranean and the Middle East. The British Prime Minister and most of his advisers were appalled. They likened Nasser to Hitler and Mussolini, and so slid inexorably into seeing any concession to him as equivalent to yet another dishonorable act of appeasement. The French were also inflamed. They too had a financial stake in the canal. They deplored Nasser's pan-Arab rhetoric and his willingness to supply arms and training to the rebels resisting French colonial rule in Algeria. A complex scheme was hatched in which an Israeli incursion would be followed by the arrival of British and French forces, who would occupy the Canal Zone, then invade Egypt and topple Nasser soon thereafter.

The entire venture proved to be an ignominious disaster. The Israeli maneuvers were condemned by the UN General Assembly on November 2. The Anglo-French invasion three days later outraged President Eisenhower, who had warned against it, and led Soviet Premier Nikolai Bulganin to threaten retaliation with nuclear weapons. There was a dramatic run on the pound but, to Britain's dismay, Eisenhower refused to authorize a short-term loan to shore up the currency until the invasion was called off. On November 6, 1955, the British Prime Minister, Anthony Eden, capitulated and informed a stunned Guy Mollet in Paris that he had no option but to agree to a cease-fire in Egypt.

The Suez debacle forced European governments to face up to the fact that they were now no more than medium-size powers, and that the United States would dissociate itself from their actions if it disapproved of them. A new resolve was born. Mollet was discussing the Common Market and Euratom treaties with Adenauer when Eden told him that the invasion was off. When he returned to the room to tell Adenauer what had happened, the latter reportedly responded by saying "Now is the time to build Europe."[2] Together they resolved the major outstanding differences between them over Euratom. The Federal Republic of Germany and the other member states accepted that France could both be in Euratom and have a weapons program, a concession that was "unfortunate but politically indispensable" to securing the agreement of France's moderate Right to both treaties.[3] The US did not like this, but dared not oppose it granted that the other five members states of Euratom did not publicly object, and for fear of alienating the French and sinking the whole project.[4] German industrialists accepted that Euratom could have a monopoly of nuclear fuel unless the European Community could not supply what was needed, in which case member states could purchase it from third parties.[5] This was also difficult for the US to swallow. The AEC owned all special nuclear material and expected other governments that received such material via bilateral agreements to do the same. In this case, it could only insist that Euratom own all the fissile materials *provided by the US* and that such material be used only for peaceful purposes, hoping to persuade other suppliers (Canada and South Africa) to impose the same constraints.[6]

Both of these concessions whittled away at Euratom's supranationality, transferring important powers back to member states. Indeed, to ensure that the treaty was signed its supporters removed contentious issues from the draft text leaving it to Euratom officials to work out unresolved issues later as they emerged. The expectation was that Euratom would eventually evolve into a supranational organization. But as France, the FRG, and Italy persisted in requesting bilateral research and power agreements early in 1957, before the treaty was signed, one embassy officer wondered if the "Euratom watering down process had not gone so far as to make it more likely there would be little real difference between Euratom and OEEC approach to Nuclear Energy Organization."[7] Dulles felt the need to remind Spaak in January of 1957 that the "US offer to cooperate liberally on a multilateral basis was conditioned on the Europeans creating an integrated Community with effective common authority and responsibility."[8] It was hoped that the "sheer size of the task of atomic energy development which must be assumed by the Community [would] force it to assume supranational characteristics."[9] That task was given to "Three Wise Men" who were dispatched to Washington to define Euratom's first program.

The First Plan for Nuclear Power in Europe and the Increasing Indifference to It

Early in December 1956, the "Wise Men"—Louis Armand, Franz Etzel, and Francesco Giordani—began work on defining the "task of atomic energy development" that was supposed to bind Euratom's member states together. Armand was to become the first president of Euratom, Etzel was the German vice-president of the High Authority of the European Coal and Steel Community, and Giordani was president of the Italian National Research Council. The Secretary of State and the chairman of the AEC quickly invited them to Washington for a twelve-day official visit in February. They were accompanied by Max Kohnstamm, who was a member of Monnet's Action Committee for a United States of Europe and who now also served as Secretary General of the "Wise Men's" staff and a major contributor to the drafting of their report. The visitors presented their plans to American industrialists, to AEC commissioners, and to Strauss, Dulles, and Eisenhower. They also visited several American nuclear installations.

The main aims of the visit were to establish the feasibility of a crash program to establish a nuclear power program on the Continent and to gain assurances that the US could provide material help. The "Wise Men" explained that, as European economies grew, their need for imported energy would rise from 100 million tons (coal equivalent) in 1957 to almost 300 million tons by 1975, corresponding to an import bill tripling from $2 billion to $6 billion in twenty years. Imports could be capped at an acceptable level of 150 million tons of coal equivalent in 1962 if Europe could install 3 million kilowatts per year of nuclear power as of 1963, for five consecutive years—a total of 15 million kilowatts by 1967. To achieve that ambitious objective, they sought "a close and enduring partnership" with the US. The US "would ensure some part of the supply of fissile materials and some of the industrial resources essential for our rapid development." Europe, in turn, would do all it could "to advance the technology and so reduce the costs of atomic power" (costs that were prohibitive relative to the price of electricity generated by coal-fired plants in the US), passing the results on to their partners. Both would also benefit politically. "Lessening the Middle East's practical hold on European industry's lifeblood would … tend to limit the risks the United States has herself accepted in that area" and would remove a "continuing threat to world peace."[10]

The "Wise Men" were given enthusiastic support by the president and the Secretary of State. Eisenhower waxed lyrical about the value of a project that was "to the benefit of the United States, of the Atlantic community, and of all the world" and secured agreement from Lewis Strauss that "a very great deal" of the fuel needed could be provided by the US.[11]

Dulles was excited about the ambitious nature of the program. It showed, he said, that Europeans were no longer satisfying their ambitions by "dominating others" (a thinly veiled reference to the Suez invasion) but rather through "constructive channels" that would help Europe, "so long the cradle of Western civilization," to "survive and grow stronger." Lewis Strauss was more circumspect but was willing to put a small group of qualified technical personnel at the disposal of the team to help them complete their report.[12]

Armand and his colleagues met with a dozen American industrialists to help define the technical details of their proposal. The businessmen injected a cold note of realism into the planning. Bob Schaetzel summed up the meeting. He regretted that, in the exchanges with the American business community, "inherent difficulties of the field were always in the forefront—the fact that there is almost no operating experience; the unexplored questions, such as the actual cost of reprocessing, will the plutonium cycle work, the actual cost of waste disposal, the megawatt days that can be estimated for fuel life; and such major industrial bottlenecks as the ability to construct pressure vessels." The "Wise Men" wanted reliable cost data. Instead they were presented with a "bewildering variety" of cost estimates. Not one businessman was willing to quote a firm figure for one of his reactors. Armand and his colleagues went away persuaded that they should start with "conventional and tested systems" if they wanted to produce 3 million kilowatts by 1962, increasing production by the same amount each year for four more years.[13] It was not clear how this building program would be balanced with the need for "technological innovation and experimentation with untested but attractive reactor systems." This was supposed to provide the added value that made Euratom attractive to the US. Without it, Europe would not be of much use to its American partners, Armand suggested.[14]

The final report of the "Three Wise Men," titled A Target for Euratom, was released on May 7, 1957, about six weeks after the signing of the Treaties of Rome. It confirmed their conservative approach. They noted that at least a dozen prototype reactors were in advanced states of design or under construction. Only two were ready for commercial use.[15] To get off to a flying start, they chose to buy these from the US and the UK or to build them under license. The British had already commercialized the gas-cooled, natural-uranium reactor, known as Pippa, that was used to generate electricity at Calder Hall. The United States' forte in reactor technology was in pressurized-water reactors (PWRs) and boiling-water reactors (BWRs), which used slightly enriched uranium. There was little enthusiasm for experimentation. The report simply suggested that other types appropriate for Europe, but not yet commercialized, would not be ignored, notably a version of the British reactor that used slightly enriched uranium and a Canadian reactor that used heavy water as a coolant. The report gratefully acknowledged assurances from the UK, the US, and Canada for technical assistance and for the fuel needed for the program. The goal was to develop a nuclear power plant that would produce energy costing 11–14 mills per kilowatt-hour (1 mill = $0.001), roughly the cost of energy produced by fossil-fuel-burning power plants in Europe.

Euratom would not produce power reactors itself. Its role was to encourage and support the efforts of national governments and industrial groups and to coordinate the atomic energy effort among the Six. This would help avoid duplication, spread development capital over a wide range of reactor types, and make more efficient use of the limited number of skilled personnel in the member states.[16] The conservative approach taken by the "Wise Men" did not inspire these political and industrial ambitions, however. In fact the Euratom Treaty was signed notwithstanding the alarming "degree of uninterest of many industrial and technical people and the general lethargy about the need for a large-scale, common program."[17]

Euratom could also consider constructing a European isotope-enrichment plant. Indeed, one of the earliest internal proposals made by the Department of State was that the US government should provide know-how, blueprints, and technical assistance, as well as financial support, for such a plant.[18] The Foreign Ministers of the Six endorsed the idea in May of 1956.[19] They established a Study Syndicate to formulate definite plans. Progress was slow until December, when the French took the bull by the horns. They formally requested that Euratom commit itself to securing independence in enriched uranium, emphasizing that this was primarily a political rather than an economic decision. In February of 1957 the Foreign Ministers obliged by declaring in favor of "the common construction of an isotopic separation plant." They invited the Study Syndicate to prepare a report in anticipation of France's decision whether to proceed, which was to be made by July of 1957.[20] The Syndicate equivocated.[21] The "Wise Men's" final report did not. After considering the possibility that Europe might build its own isotope-enrichment plant, the report rejected it, noting that it was not clear whether, in the long term, Euratom would need enriched fuel for its power reactors. More fundamentally, the "Wise Men" argued that Europe could always get fissile material from the US for one half or one third the price of manufacturing it in Europe. Indeed, in November of 1956 the AEC had announced a new price for enriched uranium: $16 per gram, a price that some thought had been set deliberately low to discourage the construction in Europe of an indigenous separation facility. In any case, some components of the gas-diffusion system (developed during the Manhattan Project) were highly classified, and the AEC made it clear that it would not make them available.[22] In May of 1957 the French announced that, while preferring a joint European effort, they would build a gas-diffusion plant at Pierrelatte on their own if they had to. No one stopped them. In fact, about a year later, in Rome, the French, German, and Italian Ministers of Defense secretly put their stamp of approval on a three-page protocol defining the cost-sharing formula for the plant at Pierrelatte that would produce enriched uranium for a French thermonuclear weapon and French nuclear submarines.

Bilateral Agreements Undermine Integration

The proposals made by the "Three Wise Men" elicited little enthusiasm. In July of 1957, a European delegation of about fifty people engaged in nuclear power was invited to the US by the AEC and American industry. The visit did nothing to dispel the technical and financial uncertainties that the report A Target for Euratom had tried to sidestep.[23] The visitors came away impressed by the financial, industrial, and manpower resources invested in the American nuclear power system. Its limits were also manifest: two major reactors, one at Argonne National Laboratory (near Chicago) and the other the Materials Testing Reactor at Oak Ridge, were down. Two others—a boiling-water reactor and a pressurized-water reactor at Shippingport, Pennsylvania—were not yet commissioned. It was clear that the UK was ahead of the US in the commercialization of nuclear power, and the visitors got the impression that American companies were still at the experimental stage. One delegate quoted by the

Belgian daily *Le Soir* came away convinced that nuclear energy was still "in the early stage of development" and that it was therefore premature for Europe to decide what kind of reactor to build. The hope that the Euratom nuclear power project would provide a supranational glue to hold the member states together was rapidly fading. National industrial and technical experts, as well as their governments, wanted bilateral agreements with the leading nuclear power producers. These would enable them to gain experience into nuclear power production, assess alternative reactor designs, and estimate costs reliably. Only then would they be willing to invest in a new energy source.

Bilateral agreements, however, were just what the proponents of Euratom feared. If not carefully handled they could both undermine an integrated European program and foster proliferation by encouraging strong national programs. Both Jean Monnet and the US Department of State were emphatic that generally no member of the Six should sign a bilateral agreement with the AEC for special nuclear material or technical support until Euratom was established. This policy was deeply resented by Franz Josef Strauss, who warned the Department of State that it would provoke a violent backlash against Euratom. To reduce the danger, the Department of State adopted a standing policy to deal with requests for bilateral agreements. It sought "assurances that the country concerned has considered fully the possible impact of bilateral negotiations with the US on the treaty-drafting work in Brussels," so as to shift responsibility for any negative consequences back onto the Europeans.[24]

France, Italy, and the FRG all sought significant bilateral agreements with the AEC toward the end of 1956. A classified bilateral agreement asked for by France in October was suspended after reconsideration of its potential effect on the drafting of the Euratom treaty, as requested by the US Department of State. The Italians asked for material for an eight-reactor program requiring between 2,500 and 5,000 kilograms of enriched U^{235} over ten years, and insisted that the United States' "refusal to proceed with an atomic power bilateral at least until the Euratom Treaty is safely born [was] a grievous blow to Italy."[25] In December, the German authorities asked for a bilateral agreement for several reactor projects and an unspecified amount of fuel. Lewis Strauss reluctantly put their requests on hold as long as there was "an early prospect of the Euratom treaty being signed," but only after a long meeting with Monnet.[26]

A compromise was found: The AEC would sign interim bilateral agreements before Euratom came into existence, but only for well-defined nuclear power projects. The rights and obligations specified in those agreements would be transferred to Euratom when it came into being (i.e., once the treaty was ratified).[27] But no sooner had the treaty been signed (at the end of March 1957) than the atomic energy officials in all three countries, unwilling to await ratification, again asked for bilateral power agreements requiring 2,200 kilograms (France), 1,500 kilograms (Germany), and 4,000–5,000 kilograms (Italy) of enriched U^{235}. This caused considerable distress in Washington. Schaetzel emphasized that these requests, if satisfied, would "undercut political support in Europe for Euratom," would strengthen the vested interests of national programs (notably in France), and, "by providing the French with an amount of fuel covering their anticipated requirements for the next six-to-ten years," would have the "further effect of freeing indigenous resources for weapons development activities."[28] Dulles noted that the requests together amounted to "from 8,000 to 10,000 kilograms of U^{235}, and [were] based on the assumption that the bilateral agreements would run for a decade," i.e., well after the time Euratom was supposed be in operation. How did the governments reconcile this with their previous agreement to seek fuel only for specific interim projects, Dulles asked before capitulating.[29] By June the demands of France, Italy, and the FRG had been acceded to, at least in part. The AEC offered 2,500 kilograms of enriched uranium each to France and the FRG and 7,000 kilograms to Italy.[30] For a year, against ceaseless pressure from the three major partners and against the wishes of the chairman of the AEC, the Department of State had fought to discourage bilateral agreements with the members of the Six. At best they had retarded national efforts; they certainly had not succeeded in curbing national ambitions to the benefit of a supranational body.

The combination of long-term bilateral agreements and concerns about the technical and financial viability of nuclear power left Euratom floundering. In July of 1957, Bob Schaetzel made an admission to Walt Butterworth, then the United States' Ambassador to the European Coal and Steel Community:

> Our problem is that we are late—I would guess by about two years. The Europeans are more confident now; much of the mystery of atomic energy has been dispelled; information has been declassified; U-235 is available; and they have

> made a start at national programs. In short, US bargaining power does not have the effectiveness it once had—or the force some Americans feel it still contains.[31]

A report on the situation in the Netherlands issued in October of 1957 confirmed that a communal approach had lost its allure: "No longer does [atomic energy] seem to be regarded as what many of us tended to consider it some two years back—a critical point of coalescence for otherwise nationalistic activities." There was "an overriding interest in a national approach to both scientific and industrial aspects of atomic energy potentialities," and Euratom was "the forgotten treaty."[32]

In December of 1957, the Council of the OEEC, with strong British support, adopted the statutes of a multilateral European Nuclear Energy Agency. The ENEA came into being early in 1958 alongside Euratom. Its aim was to "further the development of the production and uses of nuclear energy for peaceful uses in Western Europe."[33] At the same time, and with added urgency, Euratom officials and their supporters in the United States were searching for new ways to get Euratom off to a flying start when it started functioning officially on January 1, 1958. Their answer to the "drift" was the "joint program." It was an implicit recognition that the US was not the world leader in civilian nuclear power and needed an injection of public money and collaboration with European partners to advance the state of the art.

The Sputnik Shocks and the US-Euratom Joint Program

In November of 1957, Max Kohnstamm, who was now the liaison between Euratom and the United States, arranged a meeting with US officials to define an agenda for Euratom. The discussion focused on new suggestions that had emerged from Washington on how best to initiate a collaborative effort in which each partner would contribute roughly equal funds to construct perhaps four "demonstration power reactors" in Europe. Who would build them? A "strong trend" in favor of British-type reactors was noted." The British had reasonably accurate estimates of operating costs, and their use of natural uranium ensured their independence of the US. The "trend" was ignored. No one was interested in promoting the British option; in fact in July of 1957 the UK Ministerial Committee on Atomic Energy had formally ruled out membership in Euratom.[34] Their absence reinforced the argument that the "extensive and active participation" of the US was essential if Euratom was to get off to a rapid start—although of course any such assistance "would necessarily have to be limited to those reactor types in which the US [was] actively interested." With the British rival thus eliminated, the meeting laid out a possible administrative, legal, and financial scenario for a joint program that would enable the "immediate construction of, say, a million kilowatts of electrical power with US-type reactors."[35] Kohnstamm agreed to prepare the first draft of a formal proposal that Euratom could eventually submit to the American authorities.

Kohnstamm's proposal made it look as if the push for the construction of American reactors came from Europe. This was obviously meant to deflect criticism that the US Department of State was defining Euratom's agenda. In fact the joint program was a response to a new question that had been exercising Schaetzel and his colleagues in the Department of State for some time: Ask not "what we can do to help the Six" but "what they can do to help us."[36]

One thing that Euratom could do was encourage collaboration with American industry to build working reactors. As Bob Schaetzel put it in a briefing paper for Under Secretary of State Christian Herter, "the US nuclear power program is not proceeding at the pace hoped for due to a lack of domestic economic incentives coupled with the exceedingly high capital cost of full-scale reactors." "As a consequence," he continued, "the US atomic energy industry is in trouble."[37] The AEC and Congress were actively seeking ways to remedy the situation, notably by devising "some means of constructing full-scale prototype reactors" to test the technical and economic characteristics of various designs. This could usefully be done in Europe not only because nuclear power might be economically viable there but also because it would help American utility companies gain access to the Continental market at the UK's expense.

The USSR's unexpected launch of the first two Sputniks in October and November of 1957 dispelled objections to embarking on a joint R&D project in a competitive commercial sector with Western Europe. James Killian, the president of the Massachusetts Institute of Technology and the first chairman of the president's Science Advisory Committee, explained: "Overnight there developed a widespread fear that the country lay at the mercy of the Russian military machine, and that our government and its military arm had abruptly lost the power to defend the homeland itself, much less to maintain US prestige and leadership in the international arena."[38] NATO took the lead at an emergency meeting of heads of government, led by President Eisenhower, in December of 1957. They decided to put stocks of nuclear warheads and intermediate-range ballistic missiles at the disposal of the Supreme Allied Commander Europe. They also took action to enhance the number of trained scientists and engineers in Europe and to encourage them to pool their talents with those of their American peers to strengthen Western science and technology.[39] A month later, in his State of the Union address, Eisenhower graphically described the new nature of the Soviet menace:

> But what makes the Soviet threat unique in history is its all-inclusiveness. Every human activity is pressed into service as a weapon of expansion. Trade, economic development, military power, arts, science, education, the whole world of ideas—all are harnessed to this same chariot of expansion. The Soviets are, in short, waging total Cold War.

Eisenhower made a special plea to Congress to follow in NATO's footsteps and "enact the necessary legislation to enable us to exchange appropriate scientific and technical information with friendly countries as part of our effort to achieve effective scientific cooperation."[40] The Cold War was now also explicitly a struggle for scientific and technological supremacy. It was a struggle the US would win by vastly expanding federal support for scientific research and development while also increasing the emphasis on scientific and technological collaboration with its European allies. These developments were foregrounded in a memo from Lewis Strauss and Christian Herter (on behalf of Dulles) to the president promoting the joint program with Euratom. "Recent demonstrations of Soviet technological competence have intensified both the need for and interest in European integration," they wrote, and had stimulated urgent searches on both sides of the Atlantic for "new types of arrangements that will bring together the rich scientific and technical resources of the US and Europe." A joint program for the production of nuclear power was just such an arrangement. As Under Secretary of State C. Douglas Dillon put it to the Joint Committee on Atomic Energy (JCAE), "Atomic energy is rightly considered a bellwether of scientific and industrial accomplishment." Rapid progress on the joint program "will lay the foundation for the kind of meaningful scientific cooperation which is indispensable to the survival of the West."[41]

President Eisenhower endorsed the US-Euratom joint program in principle at the end of January 1958. The details were worked out rapidly over the next five months so as to secure the approval of the JCAE and of Congress before Congress' summer recess—and, it was argued, before Charles de Gaulle, who would assume office as president of France in June of 1958, could derail the movement toward European integration.

The aim of the joint program was to bring into operation by 1963 about a million kilowatts of nuclear-generated power in the Six. (The "Wise Men" had hoped for 3 million.) This was to be provided by as many as six power plants of "proven types developed in the United States."[42] They would be built, owned, and operated by utility companies in the member states of Euratom. The program was estimated to cost, exclusive of fuel, $350 million. Of this, about $215 million would be put up by European sources of capital, mostly the power utilities. The balance, $135 million, would be provided by the US Export-Import Bank in the form of a long-term line of credit at 4.5 percent interest.

The US would sell to Euratom 30,000 kilograms of enriched uranium to cover the needs of the program over twenty years. Payment for the first 9,000 kilograms would be deferred. This sale required an amendment of the Atomic Energy Act, which did not make allowance for supplying fuel to a group of countries rather than to a single country. To minimize the economic risks inherent in the fuel cycle, the AEC took the unusual step of guaranteeing both ceiling costs and the useful life of the fuel elements, a provision that Lewis Strauss deemed "vital to the success of the program."

The reactor-construction program was complemented by a joint R&D program, which was initially intended to last ten years, each party providing $50 million for it for the first five years. It would be "centered on improving the performance of the reactors involved in the program, and the lowering of fuel cycle costs." Lewis Strauss emphasized that "one of the major purposes of the proposed program would be to provide industry in the United States and Europe with important engineering experience and data on capital and operating costs." Arrangements would be made to ensure the widespread dissemination of such data in industries in the Six and in the United States.

As for proliferation, Euratom "made all of the guarantees required by the Atomic Energy Act that none of the materials supplied [by the US] will be used for military purposes." It also undertook to put in place, with the help of the United States, a systems of safeguards that would be reasonably compatible with that eventually established by the International Atomic Energy Agency, which came into being in July of 1957.

The international agreement covering the joint project was transmitted to Congress by President Eisenhower on June 3, 1958. Late in July, it was vigorously promoted to the JCAE by select industrialists and administrative officials.[43] The most striking feature of their deliberations, in Schaetzel's view, was the almost complete lack of interest in the political dimension of the project. Some of the commissioners thought there was "really no substance to Euratom" anyway. Certainly they did not seem to appreciate "the problems which Euratom has as a new institution, fighting for its life against national states." After nearly three weeks of examination by the JCAE, the Euratom Cooperation Act of 1958 was signed into law by President Eisenhower on August 29, 1958, just before Congress adjourned for the summer recess.

The US-Euratom joint program was supposed to contribute to European integration and to get Euratom off to a flying start. It pooled scientific, technical, and industrial resources on both sides of the Atlantic to meet the Soviet challenge to the Free World's technological supremacy. It satisfied Congress' determination to support the domestic nuclear industry. It was expected to provide a two-way flow of knowledge on how to build competitive US nuclear power reactors for foreign markets. It would also train manpower, consolidate the Atlantic alliance, and provide new opportunities in Europe for Westinghouse, General Electric, and other American companies. Indeed, as we shall see, by 1958 the joint program had become so central to Euratom's mission that the State Department confidently deflected a hostile attack on it from London. By contrast, Euratom's role in curbing proliferation was increasingly irrelevant. Neither Schaetzel nor the IAEA succeeded in imposing the type of controls they sought; the JCAE, for its part, was largely indifferent to the issue.

Brushing Aside the UK's Objections to Subsidies

The Department of State decided from the outset that it would negotiate the terms of the joint program without consulting the UK (though it did discuss safeguard agreements with the UK and with Canada). Bob Schaetzel characterized the attitude of the British toward Euratom as one of "absent-minded hostility" coupled with active support for the "rival"

OEEC project.[44] The UK authorities would oppose the program anyway, Schaetzel thought, on both political and commercial grounds. The British, for their part, were up in arms as soon as they learned of the terms of the fuel agreement accepted by Congress. They insisted that the proposed financial arrangements (deferred payment) and guarantees for the fuel supply (on ceiling costs and fuel life) were "something entirely outside normal commercial practice." They would "distort the normal commercial flow of trade in reactors," and constituted just the kind of export subsidy which the US had been criticizing in negotiations concerning the General Agreement on Tariffs and Trade.[45] The policy was "bound to be damaging to the prospects of the British nuclear industry in its endeavours to sell atomic power plants in Europe on a purely commercial basis."[46]

Britain's disadvantaged position was brought home when the French decided not to sign an agreement with the UKAEA, citing the more favorable terms it could get under the US-Euratom agreement,[47] The Department of State replied bluntly to objections from the UK: "The United States considers that neither the purpose nor the effect of the US-Euratom program, of which the fuel element guarantees are a part, constitute a subsidy race in reactor exports between our two governments." Listing the several areas in which the US was providing financial support to the joint program, the Department of State argued that "programs of this sweep, purpose and effect appear to the United States to be quite different from the export subsidies which we both seek to avoid in normal commercial dealings."[48] British officials decided not to pursue the matter, recognizing that they too subsidized their industry and asking only that the AEC restrict the practice to this one case.

The United States' hard line with Britain is not surprising. For months the JCAE had been asking the UKAEA to provide information on its civilian nuclear power reactors in return for information on the reactor technology used in US nuclear submarines.[49] The British had successfully resisted that demand for fear that the knowledge would be passed on to industries in the US that would then challenge UK firms' commercial prospects in the nuclear power market. By shutting the British out of the negotiations with Euratom, and by establishing a beachhead for American nuclear power reactors in Europe, the US proponents of the joint program achieved their objectives by other means, embedding them in the promotion of European integration and the consolidation of the Atlantic alliance.

Controlling Proliferation: "Enveloping" France

Euratom, as originally conceived, was intended to develop nuclear power on the Continent, to contribute to European integration, and to control the proliferation of nuclear weapons. Each of these goals had to be reassessed as negotiations over the precise form that it would take got under way. The uncertainty regarding nuclear power, the pressure for national programs, and France's determination to produce its own nuclear weapons had to be accommodated.

Early in 1957, Bob Schaetzel still hoped that Euratom could deal a near-fatal blow to the French national program. In his view, there were no good security grounds for France's having its own nuclear weapons. France's quest, as Schaetzel saw it, was "motivated by a desire to recover status in the world which they feel, rightly or wrongly, is a function of a nuclear weapons production capability." One way of deflecting them from their program was "not to hit the French directly but rather to envelop them."[50] Schaetzel hoped that Euratom could achieve that by "entangling" the French in a major European project that would drain resources from their weapons program and ultimately snuff it out. As he explained to Max Isenbergh, a special attaché for atomic energy in the American embassy in Paris:

> If a large-scale Euratom power program emerges, then there should ensue a general pattern of common projects with the scientists and technicians of the Community working together on components of the overall program. This pattern of common activity should make increasingly difficult the disengagement of the scientific and technical people of one country to work on separate, national military projects.[51]

The original crash program proposed by the "Wise Men," Schaetzel added, might "tax to the utmost the industrial and technical resources" of the Six. Paris would have to choose whether to invest scarce resources in a civilian supranational program or in a national weapons program. That dilemma would enable those in France who were not enthusiastic about a weapons program to argue that military uses of atomic energy should be held in abeyance while they concentrated on peaceful uses.

When he wrote this in March of 1957, Bob Schaetzel was counting on the proposals by the "Wise Men" to energize enthusiasm for a Community nuclear program. He still hoped that the lack of interest in the project that reflected the state of the technology (and the difficulty of getting accurate cost estimates) would gradually give way to a more positive approach. The visit of European businessmen in July of 1957 dispelled that idea. As we saw, the Europeans had the impression that the US was still in the developmental phase of nuclear power and, while not hostile to Euratom, were more concerned to strengthen their national programs. A fundamental premise of the American project, and of its capacity to shape European institutions along supranational lines by sharing technology, had evaporated.

This is not to say that the entire exercise was now fruitless. Euratom still served its purpose as a regional rather than national "solution" to a West European nuclear power program, it demanded a greater dilution of sovereignty than the OEEC, and, even if not "fully" supranational, it functioned rather like a federal system as regards the national programs. One official speculated that its right of ownership of special nuclear materials might "provide one of the few effective means for controlling or at least accounting for the imminent nuclear weapons production activities of the French and the less immediate but probably inevitable activities of the Germans."[52] Euratom also developed an elaborate system of safeguards and controls that respected its civilian character and allowed the French both to remain members and to acquire a nuclear weapon.

The US-Euratom agreement stipulated that all nuclear material produced in or imported into the Community would be owned and controlled by Euratom.[53] This applied both to enriched uranium obtained through a bilateral agreement with the AEC and to, say, plutonium produced in a French national reactor for France's weapons program. However, at the point at which that plutonium was fashioned into a weapon the material would come under the arms-control arrangements of the Western European Union (WEU).[54] This provided the legal framework for an unbroken chain of accountability for special nuclear material within the Community. And it went beyond the limited scope of the safeguards that the AEC imposed in most of its bilateral agreements having to do with nuclear power. Those safeguards covered only the material supplied by the US, and then only until the agreement lapsed.[55] The Euratom safeguards covered all nuclear materials, without any terminal date. Euratom would not "envelop" and "entangle" France in a supranational civilian power project in the ways Schaetzel had hoped, but its system of safeguards and controls could at least monitor the scale and scope of France's military program and put a brake on any attempt by Germany to develop nuclear weapons.

France's avowed intention to have an atomic weapon, the need to reduce costs, and the political delicacy of the issue led senior French officials to meet secretly with their counterparts, first in Germany and later in Italy, to discuss joint action on this issue. Three top-secret meetings were held between January of 1957 and April of 1958 while the Euratom treaty was being drawn up, adopted in Rome, and ratified.[56] This so-called FIG group adopted what the historian Georges-Henri Soutou calls a "broad agreement that would produce a community dedicated to the research and production of weapons, including nuclear weapons, guided by strategic goals that were 'harmonized' between them on the basis of a shared understanding of what was needed."[57] France also secured an engagement from its two partners to share in the cost of a gas-diffusion plant for isotope enrichment that had been discussed by the Study Syndicate set up by the Six and rejected by the three "Wise Men." Italy would pay 10 percent of the cost. The balance of 90 percent would be split equally between the host country and Germany. The FIG discussions were driven primarily by a growing fear that the United States' security guarantees were no longer credible—Europe had been "abandoned" at Suez, Britain was strengthening its independent deterrent with hydrogen bombs and was planning to withdraw its forces from Germany, and disturbing remarks made after Sputnik suggested that the US would seek to fight a "limited war" on European soil. These developments led Europeans to conclude that they had to take responsibility for their own defense. For most of the negotiators in the FIG discussions, the aim of the joint weapons program was to put pressure on the US to give Europeans a greater say over planning inside NATO. For some Gaullists it was more than that. It was a defiant move to chart Europe's nuclear-weapons programs independent of American control. A European body integrated around a civilian nuclear power program could not effectively curb these weapons developments, nor did Euratom seek to impede them. In fact it was Charles de Gaulle, not Euratom, that put a halt to the FIG project. He did so almost immediately after he assumed the presidency of France in June of 1958.

Controlling Proliferation: "Self-Inspection" and the Dispute with the IAEA

Euratom's safeguard system was based on the principle of "self-inspection." This was a major concession. It will be remembered that in May of 1956 John Foster Dulles had sought to win over the members of the Six to a supranational "Euratom" by promising that "the United States Government could make available substantially greater resources and adopt an attitude of substantially greater liberality toward a real integrated community possessing effective common responsibility and authority than would be possible for countries separately."[58] One such incentive was the right to "self-inspection."

Bilateral agreements generally "granted the United States full access to 'maintenance, production, and operating records, and other data necessary to accomplish accountability for all special nuclear materials,' and the right of access 'to all places and data necessary to verify compliance' with the terms of the agreement."[59] The Europeans were not prepared to accept such an arrangement in Euratom, as they made clear to American officials as early as December of 1956. They were unanimous that "some special arrangements must be worked out between the US and Euratom which will neither require inspection by the US along the lines of existing power bilateral arrangements nor by the International Atomic Energy Agency."[60] (The IAEA came into being officially six months later.)

Those "special arrangements," worked out in the spring of 1958 for inclusion in the US-Euratom memorandum of understanding, proved extraordinarily controversial; indeed they almost derailed the whole project.[61] In handing over to Euratom the right to control its special nuclear material itself, the AEC was willing to treat Euratom on a par with the UK and Canada, which were exempt from the safeguards policy built into the United States' standard bilateral agreement. This concession infuriated the IAEA, which wanted to define and implement a new international safeguards system that was as non-discriminatory as was possible.

The Department of State tried in vain to persuade the Europeans to drop their demand for self-inspection in the last round of negotiations on the US-Euratom agreement. As Bob Schaetzel explained, the principle of self-inspection went to the heart of Europe's sense of itself as a faithful ally and as a major force to be reckoned with in the defense of the Free World:

> A primary motivation of all the European communities has been to attain a position of political and economic equality with the United States and the Soviet Union. In view of this driving force any arrangement which places any of the communities in a subordinate or what might be called a 'colonial' status vis-à-vis the US or particularly the UK is incompatible with this primary objective of the Europeans and therefore unacceptable to them.[62]

For Euratom, self-inspection was at once an affirmation of regional sovereignty, a sign of the importance the United States attached to its success, and a marker that differentiated Europe from the rest of the non-nuclear or not-yet-nuclear powers. In adopting this stance, the Six, of course, simply affirmed their allegiance to a form of "nuclear apartheid" that secured their privileges even as they lost their colonial possessions.[63]

The principle of self-inspection was challenged by British and American officials who were looking to establish the safeguards system for the nascent IAEA. The Department of State's preferential treatment for Euratom, one American IAEA official wrote, was driven by expediency and would create only cynicism in the "under developed countries who only a few months ago in New York were solemnly exhorted by the United States as to the overriding necessity of international verification."[64] Sterling Cole, the American Director General of the IAEA, took the exceptional step of making a personal appeal to his president. Granting self-inspection to Euratom, he wrote, was likely to strike a serious if not fatal blow to the success of the IAEA, leading to pleas for special treatment from an "Asiatom" or an "Africatom" or a "Latinotom."[65] The British supported him, pointing out that self-inspection "would be used by the Soviets to confirm their propaganda thesis that major Western powers draw a distinction in practice between white, colonial states of Western Europe (which as here are allowed to police themselves) and colored countries of the near East, Africa and Asia, which cannot be trusted to do so and must be subjected to internal inspection and controls which actually are a subtle means of maintaining colonial domination and penetration of industrial secrets."[66] Max Kohnstamm did nothing to dispel this view. What would happen, an outraged British official asked him, "if Col. Nasser, for example, should insist that this principle [of self-inspection] apply to an agreement with the UAR [United Arab Republic]"? Kohnstamm replied curtly that "the art of international politics lay in discrimination," and that he "would hope that HMG [Her Majesty's Government] would be able to discriminate between Euratom and Col. Nasser."[67]

The Department of State was not as blunt as Kohnstamm, but its arguments traded on similar sentiments of Western superiority. It appealed to the "unique characteristics" of Euratom as an emergent supranational authority. And it emphasized that it was unlikely that other regional groups would demand the right to self-inspection or that they would have the "historical, cultural, and political ties with us as is the case of the Europeans."[68]

The potential damage done to the relationship between the IAEA and the Soviet Union was also disturbing. An exception for Euratom would so dent the credibility of the IAEA that the Soviets would bypass it altogether and would sign less onerous bilateral agreements with countries in the Eastern Bloc, with uncommitted countries, and even with "Free World states." It would also increase their suspicion that Euratom was a cover to enable the FRG to get its own weapons with France's help behind the protective shield of self-inspection.[69] This did not bother Dillon. The underlying rationale for allowing Euratom self-inspection, he said, "is to be found in the central role and commitment of the six nations to the Free World's ideological, political and military opposition to international Communism." The Soviet Union had shown no interest in accepting international inspection, he said. "[O]ur European allies [were] not prepared to participate in arrangements which would have the effect of inviting Soviet nationals into the heart of Europe" as IAEA inspectors.[70]

By the end of May 1958, the text of the US-Euratom memorandum of understanding had been formally accepted by the Euratom Council of Ministers and was ready to be submitted to the JCAE and to the US Congress. In one passage, the Department of State argued that the Euratom control system, which would be set up with help from the US, was the "first concrete step on the road to truly international control since it will be effectively a system of mutual control." By its "very existence and efficacy" it would "serve as an example" and would "favor the objectives" of the IAEA's incipient control system.[71] The Department of State also insisted that this was not really self-inspection, but rather was a form of international inspection in which the Germans would inspect the French, the French would inspect the Germans, and both would be inspected by the other members of the Six. And in contrast with a bilateral agreement, self-inspection would ensure that all of the nuclear material available to the Six, not just that provided by the United States, could not be secretly diverted to a weapons program.

The IAEA would hear none of this. For the IAEA, a fundamental principle was at stake in this debate. The IAEA was the product of Eisenhower's "Atoms for Peace" initiative of five years earlier. It was an international organization set up to establish a universal system of safeguards based on the American model that, it was hoped, would be acceptable to all state parties. Its safeguards provisions were not just "pious words" (Lewis Strauss) but were intended to provide "long-term and safe foreign markets for our nuclear materials and technology" (Dulles) by discouraging and containing the spread of nuclear weapons.[72] The US therefore had a particular responsibility to behave consistently in the formative period of the IAEA's safeguard system. It was not doing so. The US could not insist on "independently verified inspection for some groups of nations in the IAEA while agreeing to 'self-inspection' for others in Euratom." Director General Sterling Cole argued that this exception would be "a shattering—perhaps even a fatal—blow to the IAEA." Either alternative would be a serious blow to US foreign policy.[73]

Sterling Cole was not a bland bureaucrat. He had been persuaded by Lewis Strauss to leave his political position as Republican representative from New York State to become the first Director General of the IAEA, believing that he could play a meaningful role in removing the scourge of nuclear weapons from the world. He would not see his vision for the IAEA subverted so soon. He was sure, he wrote Strauss in May, that he too saw the IAEA as being "for the two-fold purpose of bringing to itself control of all fissionable material in the world and insuring the use of that fissionable material exclusively for the benefit of mankind through peaceful applications." Euratom was "a betrayal of that concept if it carrie[d] independent safeguard and inspection machinery."[74] Cole followed this up three days later with a personal letter to the president in which he insisted "that the safeguards or accountability aspects of Euratom be assigned to the Agency."[75]

The Department of State was deeply distressed by this development. It had prepared the ground for the JCAE and Congress to make the necessary changes to the Atomic Energy Act and to endorse the US-Euratom Agreement before the current session adjourned in August. Failure to immediately complete the agreement in which the US waived its *de jure* right to inspection would be fatal to Euratom. Late in May, Strauss, whose personal sympathies lay with Cole, begged "Stub" (as he affectionately called him) to let him know as quickly as possible if his opposition to self-inspection had "in any way softened ... so that I can be agreeable to those who find me otherwise."[76] A one-line telegram from Vienna confirmed that Cole had dug in his heels.[77] A week later he was in Washington to meet with Strauss. Together they reiterated their objections to self-inspection at a high-level meeting with officials in the Department of State.

The internal disagreement in the US government was leaked to the *New York Times* and was disclosed in an article published on June 8 under the title "A.E.C. Raises Objections to Accord with Euratom." A shorter article followed the next day.[78] Christian Herter immediately called together Cole, Strauss, Kohnstamm, and other officials and demanded that they come up with a mutually agreeable form of words that would satisfy both Euratom and the IAEA.[79] A "letter" was drawn up and initialed by those present on June 12 to be included with the draft US-Euratom Agreement. President Eisenhower himself approved the documents on June 17.

Article XI of the existing agreement stated that there would be "frequent consultations and exchanges of visits" between the United States and Euratom to ensure that the latter's "safeguards and control system effectively meets the responsibility and principles" for the peaceful use of nuclear material.[80] An elaboration of this text emerging from the exchanges arranged by Herter specified that this included "permission by each Party for the other Party to verify by mutually approved scientific methods" such "effectiveness."[81] These last-minute discussions also secured an agreement that if the IAEA were to establish an international safeguards and control system, the US and Euratom would "consult regarding the assumption by that Agency of the safeguard and control over the fissionable material utilized or produced" in the foreseen joint program. The earlier version had simply committed the partners in the joint program to ask the IAEA for assistance with their system of safeguards and controls if they felt a need for it.[82]

"Our right to inspect is clearly established in the documents initialed last night"[83]—thus did Eisenhower seek to reassure Bernard Baruch, the man who had spearheaded the first attempt to control nuclear proliferation through the United Nations, and who was now deeply concerned about granting Euratom self-inspection. The president blurred the issue to assuage Baruch's concerns. The US did not in fact gain the *de jure* right to inspect, but only the more limited right to verify that Euratom was "effectively" implementing safeguards and controls and was using scientific methods to do so. Such mutually-agreed-upon non-intrusive safeguards, as they are now called, specifically excluded non-Euratom inspectors from entering nuclear facilities unless invited. As the British pointed out, American officials would be "allowed to see the appropriate Euratom records and to visit installations on request, and vice versa." But—and here was the rub—"both sides would however do so as 'guests,'" respecting Euratom's right to run its own inspection system.[84] The official historians of the AEC, Richard Hewlett and Jack Holl, concur. Euratom "maintained its refusal to submit to an externally administered safeguard system." The US, for its part, "had been obliged to accept a system that included the right to audit but whose ultimate sanction would simply allow the United States to terminate the cooperative program" if it was not satisfied with Euratom's capacity to safeguard material.[85]

Eisenhower's willingness to override Cole highlighted the most important consideration that had driven US policy on Euratom. In a personal letter to the Director General, the president put it as follows: "Both the Legislative and the Executive branches of this Government have given consistent support to the European integration movement over the last ten years. Euratom is a new and important element of this movement. We wish it to succeed."[86] In subsequent hearings before the JCAE the issue of safeguards and self-inspection was barely raised. Indeed, Senator John Pastore, the chairman of the JCAE subcommittee that considered the Euratom program, apparently told Max Kohnstamm that the IAEA would "have to face the fact that its task within the next years will be largely educational," and that it would be essentially irrelevant to the relationship between Euratom and the United States.[87] With no strong support in Washington, Cole could not but yield.

Bob Schaetzel was disappointed by the JCAE hearings on Euratom. They were dominated by a previous breakdown of trust between the Joint Committee and the AEC that distracted attention from an in-depth debate on the joint program. The proponents of Euratom were put on notice that they could not count on the unqualified support of their American partner, and that the project would be judged by its results. In short, as Schaetzel put it ironically after it was all over, "Now all we have to do is organize the

Program and get those damn reactors built."[88] Doing so proved far more difficult than Schaetzel had imagined. Walt Butterworth, now US Ambassador to the European Communities, spelled out the "Major Problems of Euratom-US Relations" in a twelve-page survey for the Department of State and the AEC in February of 1961.[89] Fifty joint R&D projects worth $11 million were under way, and US firms were involved in almost half of them. However, to date only one firm proposal for a power reactor had been submitted for support under the US-Euratom joint program.[90] This was due primarily to the "changed energy situation, in which the wide availability of coal and oil and increasing availability of natural gas diminished the early need for nuclear power." European industrialists were loath to invest heavily in large-scale nuclear power plants, which they feared might not be economical. The use of the loan from the Export-Import Bank ran afoul of the "Buy American" clause in the contract, contrary to assurances that it would not be enforced strictly, and that funds borrowed from the Bank could be used for purchases of nuclear equipment in non-US markets.[91] A final blow was delivered by the new administration of John F. Kennedy, which denounced collaboration with Euratom as a drain on American technology that provided little in return.[92]

The Euratom that existed in 1960 was a far cry from the institution imagined by Monnet and Spaak just five years earlier. Historically, though, the importance of a European atomic pool as a cement holding an integrated structure in place as the European *relance* gained momentum cannot be denied. Euratom was a beacon of hope on the difficult path to the Common Market, its flame kept alive by American promises of sharing knowledge and materials in the nuclear domain. Its situation was always fragile, however, its momentum sustained by external and sometimes unexpected developments. The utopian dreams inspired by "the peaceful atom" fueled the project in 1955. The Suez debacle re-energized the flagging enthusiasm for European integration at the end of 1956. The launch of Sputnik in October of 1957 persuaded a hesitant Congress to endorse an Agreement for a collaborative R&D project between the AEC and Euratom. And just as this external nourishment kept Euratom buoyant at first, so did its fortunes recede when it was withdrawn. Nuclear power never lived up to its commercial promise. New, relatively inexpensive alternative sources of energy were discovered. Political support waned. The Euratom bubble burst, and the organization settled down to a more modest existence on the periphery of the European Economic Community, which it had helped bring into being in 1957.

Concluding Remarks: Integration, Proliferation, and Soft Power

The web of connections between American soft power, technological leadership in atomic energy, European integration, and nuclear non-proliferation provided the leitmotif for this chapter and the preceding one. What conclusions can we draw from these two chapters?

First, we need to step carefully before celebrating Euratom as an instrument of European integration. For Monnet and Spaak, integration was synonymous with the establishment of a supranational organization with centralized powers over most aspects of nuclear power production, from the provision of fissile material to ensuring that it was not diverted to military ends. This vision for Euratom could not withstand the centrifugal pull of the French and German national agendas in 1956. Euratom's supranationality was diluted to accommodate France's determination to build nuclear weapons and German industrialists' demand for greater control over fissile material. It suffered a further setback in 1957 when European and American industrialists evaluated the prospects for nuclear power and found them to be technologically and financially uncertain. Many people believed that governments first had to build up a better understanding of the potential of atomic energy in national programs before they collaborated intimately with others. This priority was expressed in the pressure that the Six put on the AEC for bilateral agreements in 1957, even after the Euratom treaty was signed in March of that year. It can also be inferred from the relative success of the R&D leg of the joint program and the lukewarm response to building prototype reactors. Eventually Euratom was an "integrated" organization in that it provided a forum for the shared discussion and coordinated implementation of select nuclear power policies and research projects by its member states, which collectively funded its activities. It also tied the FRG into a multinational organization established by treaty—the definition of integration that sustained Eisenhower's and Dulles' enthusiasm for the projects that came into being with the Treaties of Rome.

Second, Euratom's ability to curb proliferation was limited. Schaetzel's hope of "enveloping" France in the organization and draining resources away from a national military into a supranational civilian atomic energy project was stillborn. In fact, accommodating the French weapons program actually heightened the risk of proliferation, as no one saw how Germany could legitimately be stopped from acquiring nuclear weapons if France had them. Indeed, Adenauer remarked in October of 1956 that Euratom might provide a way for the FRG "to produce nuclear weapons herself."[93] And as one US official explained, the safeguards system as eventually agreed did little more than "account for" rather than "control" production of nuclear weapons by France and, ultimately, Germany.

Gunnar Skogmar has tried to salvage Euratom's historical significance by arguing that the organization successfully prevented the manufacture of atomic weapons in Germany, if not in France.[94] He overstates the success of Euratom as an instrument of non-proliferation. If security studies teach us one lesson, it is that the decision to embark on a weapons program is a political one, and that many factors working in unison prevent states from acquiring nuclear weapons.[95] The United States' promotion of Euratom was driven by Eisenhower's dedication to spreading the benefits of "the peaceful atom" and by his administration's commitment to European integration both as an antidote to nationalist rivalry and (as Skogmar puts it) as an instrument to "prevent, retard or minimize nationally controlled weapons in Germany and France" or the development of a third force outside of NATO.[96] The goal of prevention was soon dropped. Indeed, not only did Secretary of State Dulles relinquish the goal of non-proliferation; he also took the initiative in suggesting that France simply delay its first nuclear test by four years to respect Euratom's peaceful mission. Retarding rather than stopping proliferation was the best Dulles could hope for. As he told the British Foreign Secretary in March of 1957, he was reluctant to sever all ties with the French weapons program, hoping that the attraction of technological collaboration could be used to slow it down. In sum, the most that can be said for Euratom as serving the goal of non-proliferation—and that is achievement enough—is that membership posed severe institutional and material constraints on any aspirations for nuclear weapons that Germany, may have had. It could not however, stop the construction of a parallel nuclear-weapons program, such as the one that was discussed secretly by France, Germany, and Italy for about 18 months before de Gaulle put a stop to it in June of 1958.

Grégoire Mallard does not share Skogmar's view that Euratom curbed proliferation. On the contrary, he feels that its civilian program served as a cover for a weapons program.[97] His rich analysis of the agreements formalizing the relationship between Euratom and the US focuses on ambiguous language that, he says, misled the JCAE and Congress as to the military intentions of Euratom's member states. Mallard contrasts the public civilian nuclear power "charade presented by the Three Wise Men" with the "opacity" ("produced by a series of lies") with which the US Executive Branch and the Europeans "mask[ed] the advancement of their planned nuclear weapons program from the US Congress."[98]

There is no doubt that certain clauses in the aforementioned agreements—for example, those on safeguards—were ambiguous if only because the Euratom agreements had to be phrased to allow France to develop nuclear weapons. But Mallard's fascination with "the duplicity at work in the art of diplomacy" has misled him.[99] The enormous effort that was put into launching the joint civilian nuclear power program in Europe and the resentment its "subsidies" caused in the UK surely disqualify it as a charade. Nor did the proponents of the US-Euratom agreements attempt to "mask" the military aspects of the organization from the JCAE, as Mallard implies. Christian Herter, testifying before the JCAE in July of 1958, said explicitly that "adherence to the Treaty by the member states does not preclude a member itself from embarking on a nuclear weapons program."[100] Of course, there was no public mention made of the secret FIG negotiations—why should there have been, since they were embarked on quite separately from the Euratom negotiations? In any event, it is not certain that the JCAE would have objected. As one briefing document noted, there was "no conflict" between "the peaceful nature" of Euratom and France's intention to pursue a program to develop nuclear weapons—"certainly no more conflict than in the United States undertaking peaceful uses programs while continuing to strengthen our military potential."[101]

When the JCAE hearings were over, Bob Schaetzel wrote to Walt Butterfield summarizing the proceedings. What struck him was that the JCAE "appeared to care little about the contribution the Program promises to the cause of European integration." "There was," he continued, "no interest whatsoever shown in the French nuclear weapons program, the relationship of the Euratom program to it or where the Germans may eventually go into this field." Schaetzel went on to remark that in some "side discussions" the consensus of the JCAE seemed to be "that if they were in the shoes of the French they too would be pushing ahead with nuclear weapons development."[102] There was banalization of the military dimension, not opposition. Put differently, and contra Skogmar and Mallard, no one on the JCAE seems to have even *expected* Euratom to serve as an instrument of non-proliferation.

The early history of Euratom teaches us a lot about the exercise of soft power.[103] All the ingredients that constitute Nye's concept were seemingly present when Euratom was launched in 1955: an asymmetry in knowledge/power in favor of the United States, co-option rather than coercion into a joint project, and the determination to shape institutions along lines that the "dominant state prefers"—i.e., a supranational Euratom rather than a looser OEEC-like collaboration. Schaetzel was brutally clear in July of 1957 as to why this structure crumbled: "Our problem is that we are late—I would guess by about two years." In fact it gradually dawned on him that the American atomic industry was "in trouble." The "tremendous overall engineering knowledge, personnel depth etc." in the US atomic energy program did *not* translate into a "three year head start" in the civilian applications of nuclear power. When it came to supplying nuclear power plants, America's bargaining power was far less than had been imagined; they had no chips.

The heart of the problem lay elsewhere, however. The coherent exercise of the one soft-power asset the US had—seemingly limitless amounts of fissile material—was thwarted by internal disagreements between the Department of State and the Atomic Energy Commission. The success of the pro-integration policy promoted by Eisenhower, Dulles, and the "federalists" in the Department of State pivoted on the Six not seeking bilateral agreements with the AEC, and the AEC refusing them if they did. That did not happen. Atomic energy establishments in the Six were determined to develop national programs, and the AEC was happy to oblige. The Department of State's soft power was heavily circumscribed because the political weapon that it needed to steer the Europeans along its preferred path was in the hands of the AEC. As a result, it had to content itself with a diluted mode of integration that was compatible with the technological, industrial, commercial, and political realties of nuclear power in Continental Europe at the time.

3

"A Substantial Sop," or "Positive Disarmament"? Johnson, Erhard, and Bilateral Space Collaboration

Late in December of 1965, Chancellor Ludwig Erhard of the Federal Republic of Germany made a brief trip to Washington. Two contentious issues dominated the agenda of his discussions with President Lyndon Johnson: the role of the FRG in a multilateral force (MLF) equipped with nuclear weapons and the "offset" costs to Germany of keeping American forces in the country. The confrontational tone changed at an official banquet for Erhard in the State Dining Room on the evening of December 20. In a toast to his distinguished guest, Johnson emphasized the importance of the FRG in consolidating the Atlantic partnership. He especially thanked Erhard for his support for the war in Vietnam. Johnson said that he looked forward to the United States' learning how Germany had reduced air pollution in its cities. And he announced a major new initiative in space science. A bilateral agreement had already been signed between the two countries to launch the FRG's first satellite, named Azur.[1] The time was ripe to do something more ambitious—to do "together what we cannot do so well alone." Johnson mentioned a probe to the sun and a probe to Jupiter as appropriate ventures that were "very demanding" and "quite complex" and that might contribute "vastly to our mutual knowledge and our mutual skills."[2]

Erhard was back in Washington nine months later in September of 1966. He was willing to renounce a "hardware solution" to the MLF that granted Germany limited access to nuclear weapons.[3] But he was determined to reschedule the onerous offset payments, which were draining his government's coffers and his political capital. Johnson would hear none of it. And notwithstanding the brevity of Erhard's visit, that lasted just two days, he was rushed to Cape Kennedy with the president and with NASA Administrator James Webb to see for himself just what the Apollo manned lunar landing program involved. Johnson again emphasized his wish for more expansive space cooperation with Europe, and particularly with Germany.[4]

These public displays of offers to collaborate in an advanced space project are usually ignored or glossed over in the literature that concentrates on the disagreements between Johnson and Erhard over nuclear sharing and the offset arrangements. Thomas Schwartz's classic account of Johnson's relationships with Western Europe includes a photograph of a rather glum-looking Erhard at Cape Kennedy in September of 1966, but sees Erhard's trip to Florida primarily as demonstrating the technological gap between the two nations.[5] Hubert Zimmerman judges the whole episode to be of minor significance.[6] Wolfgang Finke, an official in the FRG's Ministry for Research and Technology for almost 30 years, confirmed these assessments in an email exchange with me in 2007. As he put it, "the Johnson proposal for a joint space project was mainly meant to sweeten the bitter pills Johnson had to give Erhard [as regards offset payments] and at the same time to underline the vast superiority of the United States in that important and innovative field."[7] There was no particular connection between the proposal for space collaboration and the more serious issues that divided the two leaders.

This chapter argues otherwise. It places Johnson's offer to Erhard for a major joint space project in the broader context of his administration's evolving policy on the *N*th-country problem. In particular, it argues that Johnson's offer of collaboration in space exploration must be understood as one of a variety of measures devised by his administration to curb nuclear proliferation. Those who believed that the US should do what it could to retard and if possible thwart other states' efforts to acquire nuclear weapons did not think that any one measure would constrain them. Pressure had to be applied at many different country-specific points to dissuade non-weapons states from embarking on bomb programs. Technological collaboration in a major civilian space project was deployed as a prestigious substitute for a national weapons program—a tool of "positive disarmament," as it came to be called. The administration tested the policy in India and in Japan. It also tried it in Germany. Far from being tangential to the major points of friction between Erhard and Johnson, the president's offer of space collaboration was supposed to help reduce tension over nuclear sharing (and offset payments).

Prehistory: The Chinese Nuclear Test and the Gilpatric Report

In October of 1964, the People's Republic of China (PRC) successfully tested its first nuclear weapon. The test itself did not surprise the Johnson administration, though the use of highly enriched U^{235} as fissile material did.[8] President Johnson downplayed the significance of the PRC's breakthrough, whereas others insisted that it posed a major threat to American interests in Asia. Of far greater concern to many in the Johnson administration was proliferation beyond China. India, which had lost a border war with the PRC in 1962 and which had no formal security guarantees with either the USSR or the US, was thought likely to embark on a nuclear-weapons project. Japan was another proliferation candidate, notwithstanding the hostility of the Japanese public to such a possibility; indeed, Japan's new prime minister, Eisaku Satō, responded to the Chinese test by suggesting that his country should have a nuclear deterrent. Sweden and Israel were among the other states that already had the technological capability to produce nuclear weapons. It was hard to see how the Germany could renounce nuclear weapons much longer. CIA Director John McCone immediately warned Johnson that the "Germans probably possess sufficient nuclear technology to develop weapons" and that they might exercise that option "if the political situation so dictates."[9]

Within a fortnight of the Chinese test, President Johnson had taken steps to establish a high-level committee to explore what the US might do, alone or in concert with other governments, to limit nuclear dissemination. Roswell Gilpatric, a Wall Street lawyer and a former Deputy Secretary of Defense, chaired the committee, which included George Kistiakowsky (a professor of physical chemistry at Harvard University and a former science advisor to President Eisenhower) and Herbert York (a professor of physics at the University of California at Berkeley who had been the first director of the Lawrence Livermore Laboratory). The committee's formal brief was to study "the problem of preventing the spread of nuclear weapons."[10]

The committee consulted widely with officials in the Department of State, the Department of Defense, the NSC, the AEC, and the Arms Control and Disarmament Agency.[11] It discussed a range of policy options, from assuming that proliferation was inevitable (even desirable in some cases) to elevating the prevention of nuclear dissemination to an overriding foreign-policy goal of the United States. The committee's final report (probably drafted by Spurgeon Keeny, an arms-control expert who worked for the National Security Council) was delivered in January of 1965. The committee recognized that there was a diversity of opinion within the government, and indeed among its members (at least initially), about the feasibility and the costs of preventing nuclear proliferation. It nevertheless came up with a unanimous proposal that "preventing the further spread of nuclear weapons [was] clearly in the national interest, despite the difficult decisions that [would] be required."[12]

The significance of the Gilpatric report was that it advocated sacrificing the interests of allies if that was necessary to promote nuclear non-proliferation. It insisted that the world was "fast approaching a point of no return in the prospects of controlling the spread of nuclear weapons." It acknowledged that the US had been concerned about nuclear dissemination since 1945, but demanded that non-proliferation policies be given "far greater weight and support than they have in the past" by all the agencies of government that dealt with them. And it emphasized that the US should, as "a matter of great urgency," increase the intensity and the scope of its efforts to prevent the further spread of nuclear weapons. The US was to be proactive in promoting multilateral treaty commitments to non-proliferation, a comprehensive test ban, and the creation of nuclear-free zones. More specific country-by-country measures were needed "to influence the decisions of individual non-nuclear powers not to undertake the development or acquisition of nuclear weapons and to secure workable commitments to this effect."[13] Economic restrictions were one means to that end. Another means, of particular pertinence here, was the use of scientific and technological collaboration to promote alternative sources of national prestige.

Spurgeon Keeny specifically identified technological collaboration as a political lever for curbing the nuclear aspirations of India and Japan, each of which had a strong research base and a heightened sense of vulnerability to a new, communist nuclear power nearby. At one of the very first meetings held to discuss the implications of China's test, Keeny "noted Japanese and Indian interest in finding some way to demonstrate their own technological achievement, not necessarily in the military field, in order to offset the Chinese test."[14] This strategy was incorporated in the Gilpatric Report, which suggested that India and Japan could be discouraged from going nuclear by mixing security guarantees with, inter alia, assisting India in "reasonable and economically justifiable scientific programs designed to build the prestige she might otherwise attempt to obtain from the development of a nuclear device" (including natural resources, health and birth control).[15] Much the same was suggested in regard to Japan: "We should, as in the case of India, attempt to help the Japanese with appropriate prestige alternatives."[16]

Keeny's proposal was not given any particular weight over others in the Gilpatric report; it was simply one of the instruments that the US had at its disposal for "halting or retarding the spread of nuclear weapons." Be that as it may, the idea gradually gained ground among senior officials in the administration. It was actively pursued in discussions with leaders in India and in Japan. It was also explored as part of the solution to the "German question"—that is, the fear that the FRG would seek to follow in France's footsteps and acquire an independent nuclear-weapons capability.

Positive Disarmament

Scott Sagan has emphasized the importance of prestige as an aspect of nuclear statecraft, using France's quest for an independent nuclear deterrent as an illustration.[17] American security guarantees were in place. De Gaulle's national nuclear strike force could hope to do little more than "tear the arm off the Russian bear."[18] Why persist with an expensive program that would be largely ineffective in the event of a Soviet attack? Sagan suggests that France's weapons program mattered above all because it boosted the country's *grandeur* and national prestige, and because it was the kind of technological feat that major powers had to perform to affirm their legitimacy. France's Minister for Defense (Ministre des Armées) Pierre Messmer nicely summed up the link between national prestige and an independent nuclear capability in 1963: "One is nuclear or one is negligible."[19] Gaullist France associated national pride and prestige with a nuclear capability.[20] For de Gaulle there was no other way for France to regain the influence it had lost through defeat and occupation in World War II, decolonization, and the rise of the United States to a position of world leadership.

Spectacular space projects similarly served to boost national prestige in the 1960s, when the space race turned each success achieved by one of the superpowers into a demonstration of national prowess. Secretary of Defense Robert McNamara made the point in April of 1961, just a month before President Kennedy's famous commitment to a manned lunar mission: "All large scale space programs require the mobilization of resources on a national scale. They require the development and successful application of the most advanced technologies. Dramatic achievements in space therefore symbolize the technological power and organizing capacity of a nation. It is for reasons such as these that major achievements in space contribute to national prestige."[21]

By the mid 1960s, the achievements of NASA's Gemini program were thrilling TV audiences the world over and providing tangible evidence of American technological prowess and supremacy in the conquest of space. To explore how best to use that asset, a special Working Group on Expanded International Cooperation in Space was set up by the National Aeronautics and Space Council at the request of its chairman, Vice-President Hubert H. Humphrey, in March of 1966. This inter-agency group, chaired by Herman Pollack of the Department of State, included representatives from NASA, the Department of Defense, the Atomic Energy Commission, and the Arms Control and Disarmament Agency. Senior officials were quick to see the prestige value of technological collaboration on satellites and (more prudently) rockets as alternative sources of national pride to nuclear-weapons programs. Secretary of Defense Robert McNamara insisted that NASA go beyond simply asking its partners what their interests were in a joint space effort. The US, he said, "should look at each country's technological capacity and then push them to use this capacity to the full in the space field." Developing the idea, McNamara argued that "at the present time many countries felt they should make technological progress through military hardware programs." "A divergence to a space program would be wholesome for all concerned," he continued, and could satisfy the same objective.[22] NASA Administrator James Webb clarified the tenor of the discussion in a subsequent letter to McNamara, confirming that the US ought to actively encourage "foreign involvement in space technology as a means of diverting energies away from the development of nuclear weapons systems."[23] Hubert Humphrey labeled this "positive disarmament."

A word of caution is needed at this point. In the security studies literature, "diversion" usually refers to a policy adopted by an aspirant nuclear state to clandestinely move resources and raw materials from a putatively civilian program into a potentially military program. Here the meaning is inverted. It is taken for granted that the US was dealing with major industrial powers that had the capacity, in principle, to embark on weapons programs. It is being suggested that the US, by offering technological assistance to civilian nuclear and space programs, could cripple or at least retard other countries' weapons programs by diverting scarce resources away from them, thereby stabilizing the international system under its leadership.

Webb immediately invoked "positive disarmament" in an attempt to denuclearize India. He suggested to U. Alexis Johnson, the Deputy Under Secretary for Political Affairs in the Department of State, that through collaboration with India on space projects "some Indian energies might also be diverted from concern with nuclear weapons development, the more so perhaps as the success of the experiment contributed to India's prestige in Asia."[24] In fact, space collaboration was welcomed by the Indian space lobby, which feared that available resources for major technological projects might be drained by an increasingly likely bomb project. Vikram Sarabhai, the "father" of the Indian space program and an ardent promoter of satellite TV for Indian villages, told the US Department of State in October of 1966 that "there was some pressure within India to build a nuclear bomb, and to deflect this pressure India needed to do something else to demonstrate an advanced scientific capability."[25] From these small beginnings, NASA eventually worked with Indian officials to launch a telecommunications satellite that beamed public-service programs to thousands of Indian villages in the late 1970s. As it happens, this was not at the expense of ongoing investments in a nuclear program that resulted in India's testing a bomb in 1974.

In September of 1966, the Arms Control and Disarmament Agency advocated the policy of "positive disarmament" for Japan too: "A significant program of space cooperation directed towards enhancing Japan's space capabilities can have beneficial nuclear non-proliferation activities by bolstering her prestige, demonstrating her worldwide and regional scientific prowess and affirming the US involvement with that nation."[26] Space cooperation was broadly favored in Japan, where public opinion was deeply hostile to developing nuclear weapons. A dispatch to Washington from the embassy in Tokyo in November of 1964, just a few weeks after the first Communist Chinese nuclear explosion (CCNE), indicated that "the Prime Minister would like very much to see a satellite orbited by Japan as a counter to the CCNE and as proof to the world of Japan's highly advanced scientific skill."[27] The same argument was repeated some time later by Yasuhiro Nakasone, a member of the Japanese Diet and the chairman of its Subcommittee on Space Development. In November of 1966, in the daily *Mainichi Shimbun*, Nakasone wrote: "Advanced nations today are symbolized by science and technology, particularly atomic energy and the achievements of space development." "If China should launch an artificial satellite for peaceful use into space ahead of us," he added, "the sense of hopelessness of the Japanese will be so great that no one has the heart to see it."[28]

The notion that the US could divert national resources away from nuclear-weapons programs and toward prestigious scientific and technological projects by offering to share important civilian space technology was, then, very much in the air in the mid 1960s. It was raised indirectly when Johnson met with Erhard in December of 1965.

Johnson, Erhard, and the MLF

> If it becomes apparent in your talks with Erhard that very little will come out of nuclear sharing program, you might consider discussing with him greater participation by the Germans in the space program. The Germans need an outlet for their energies.
> —George McGhee (US ambassador to the FRG) to President Lyndon B. Johnson, December 1965[29]

Johnson inherited the MLF from Eisenhower and Kennedy.[30] His predecessors had proposed establishing a multinationally owned and operated seaborne force of medium-range ballistic missiles. The aim, as Trachtenberg puts it, was "to absorb and deflect the pressure for national nuclear forces which, if unchecked, would lead sooner or later to a German nuclear capability."[31] Eisenhower saw this as evolving into an independent European nuclear force; the Kennedy administration insisted on making the use of the force subject to an American veto. As endorsed by the Kennedy administration in 1962, the MLF foresaw a fleet of nuclear-armed surface ships manned by mixed-nationality crews. Hal Brands emphasizes its prestige value: it was intended to "satisfy German status anxieties by allowing the FRG partial control of nuclear weapons … while at the same time avoiding the dangers associated with an independent German deterrent."[32] The proposal was unacceptable to the Soviet Union, which saw it as launching the FRG on the path to its own nuclear strike force. De Gaulle dismissed the MLF as a Trojan Horse for US influence in Europe. The British Chiefs of Staff regarded it as expensive and superfluous. Kennedy and then Johnson persisted all the same, pressured by the MLF "theologians" in the Department of State.

The Gilpatric report emphasized that it was urgent to "make it possible for the Soviets to take the lead with us in seeking world-wide support for a non-proliferation treaty." It recognized that this would alienate the FRG, but it insisted that that was the price of securing the USSR's assent to the treaty. This forced Johnson to reconsider his stance and to think of scrapping the MLF.[33] He was obliged to backtrack in the face of strong opposition in the US and deep hostility in Germany to any measures that might bury the plan.

Before coming to Washington in December of 1965, Erhard was strongly urged by both his own and some US foreign-policy advisers not to yield on nuclear sharing, even if his resistance would impede the brokering of a non-proliferation treaty with the USSR.[34] To emphasize that the FRG wanted to be party to discussions over the deployment of nuclear weapons in Europe, Erhard told Johnson of his sympathy for de Gaulle's view that "a nation protected by others with no say of its own could not regard itself as truly sovereign." Erhard added that it "was impossible to assume that Germany will go forever without a nuclear deterrent." He even hinted that it might soon be in a position to produce its own. Some of his scientists had told him, Erhard reported, that before too long "nuclear weapons could be produced much more cheaply and technological aspects mastered more easily, so that many smaller countries would be able to afford nuclear weapons," replacing conventional forces with them—no doubt a reference to the advanced work German firms were doing on gas-centrifuge enrichment of uranium (see chapter 5). To allay Johnson's fears, Erhard then expressed his support for a "fully integrated system" that would "prevent Germany from taking the lead."[35] At the end of their conversation he handed Johnson a paper that spelled out his views on nuclear sharing.

Erhard's paper made it clear that Germany sought "an appropriate direct share in nuclear responsibility" by being a partner in a "jointly owned and jointly administered" force equipped with nuclear weapons whose use could be vetoed by the US.[36] The force, it seemed, would include British and American submarines in equal numbers, as foreseen in British Prime Minister Harold Wilson's idea of an Atlantic Nuclear Force.[37] This was a blow to those who hoped that Erhard would be satisfied merely with a greater say in NATO decision making in what came to be called the Nuclear Planning Group (the "soft-ware solution"). Erhard still wanted a "hardware" solution to the FRG's quest for military security. He also wanted political parity with France and Britain.

The United States' ambassador to Bonn, George McGhee, had warned Johnson in advance that Erhard was unlikely to yield on the main provisions of the MLF. He suggested that "greater participation in the space program" might draw Erhard away from his policy on nuclear weapons. The Gemini flights had "greatly excited" the German public. A major space project would provide an alternative "outlet for their energies," as McGhee put it.[38] Johnson raised the point when he discussed space collaboration with Erhard. Emphasizing the major role that Wernher von Braun and other German émigrés were playing in the American space program, he read Erhard a draft of the speech he would make that night at the State Banquet, proposing a joint project in space science. Johnson added that he "thought such a cooperative effort would do more to weld the two peoples together than all the conversations they could have on nuclear submarines and nuclear deterrence."[39] It was a strong hint that the warm relations between the US and its most important Continental ally were better served by a joint space project than by endless disputes over Germany's access to nuclear weapons through the MLF.

An Additional Argument for Space Cooperation: Closing the Technological Gap

Erhard was not an enthusiastic proponent of expenditures on space programs. Indeed, in December of 1965, just before he left Germany to meet with Johnson, his Research Ministry was complaining bitterly about the tight-fisted approach of his administration. It had managed to secure long-term financial support for the FRG's first satellite only because a memorandum of understanding had been signed with NASA. By contrast, "funding for the development of a second scientific satellite and the conduct of further experiments, already planned under a specific program and agreed in preliminary talks with NASA, has so far been refused" by the Finance Ministry.[40] A major space project would be a hard sell.

Ambassador McGhee took on the challenge. Indeed, soon after Erhard's visit to Washington he devoted the entire content of a long address at the University of Bonn in January of 1966 to describing the history of America's space program, extolling the many scientific, technological, and social benefits it produced, and emphasizing the importance of international space collaboration and of US-German collaboration in particular.[41] McGhee touted the probes to Jupiter and the sun as answering questions "of basic scientific concern to all mankind." The joint project would take the present stage of US-German cooperation to a new level. Its organizational complexity would introduce Europeans to American practices of systems management. The collaboration between industries would "constitute an additional contribution to the concept of Atlantic Partnership within the Atlantic Community." Finally, the ambassador emphasized that increased national investment in the fields of science and technology opened up by space exploration was of "vital importance" not only to Germany and to Europe but also "to the technological future and overall strength of the Free World."

McGhee's lecture overlapped with the start of a week-long visit to the FRG by Donald Hornig, Johnson's Special Assistant for Science and Technology. Upon returning to Washington, Hornig wrote to tell the president that he had found widespread enthusiasm in Germany for expanded scientific and technological collaboration with the US. He identified defense and space as two fields in which the technological gap between the US and Europe was widening. As was mentioned in the introduction to this book, he concluded his short report on an upbeat note, insisting that he could not "emphasize too strongly the intellectual leadership which the pre-eminence of American science and industrial technology gives us in the world today." It could be used, Hornig suggested "as a powerful political force at little financial cost."[42] Johnson's proposal to Erhard had stimulated a broader assessment of the role of scientific and technological collaboration as an instrument of American foreign policy. It could serve as a tool of "soft power" to consolidate the Atlantic alliance and the strength of the Free World. It could also help close the technological gap between the US and Europe.

The technological gap was a major preoccupation of the Johnson administration well before it was widely publicized in Jean-Jacques Servan-Schreiber's 1967 book *Le défi américain.* In fact, Johnson himself commissioned a special report from Hornig on the topic late in 1966.[43] Months earlier, Hornig had concluded that "the basic problem of the 'technological gap' relates to such fundamental issues as the size of European markets, European management procedures, and policies toward capital investment in industry, research and development, and education, and that these problems could not be cured by any package of technological proposals that we might offer."[44]

Why did the Johnson administration care about the technological gap? It will be remembered that, in the post-Sputnik panic, Eisenhower—first in NATO in December of 1957 and then in his State of the Union address in January of 1958—had emphasized that the strength of the Free World depended on transatlantic scientific and technological collaboration. Such collaboration would be fruitless, though, unless both partners could contribute meaningfully to the production of useful knowledge, goods and services. Talk of a "technological gap" implied that Western Europe was so far behind the United States that reciprocity was impossible. Europe's research system was a dependent rather than an equal partner, and its industries were uncompetitive and overwhelmed by their high-tech American rivals. This had to be remedied. To that end, a number of areas in which closer scientific, technological, and industrial cooperation might improve matters were identified. Space exploration was one of them.

In February of 1966, Arnold Frutkin, the director of NASA's Office of International Affairs, went to Europe with a NASA team to gather support for the joint mission proposed by Johnson, now labeled the Advanced Cooperative Project. Frutkin firmly believed that a major investment in space exploration could lift the scientific and technological capacity of a nation across a broad front. He emphasized the need for increased national expenditures again and again in his discussions with Europeans. The US program, he claimed, had pushed established scientific and technical disciplines to probe new frontiers in a wide variety of fields, ranging from physics to geodesy and from materials to structures. "In fact," he insisted, "we may with increasing confidence say that the peculiar quality of space science and technology is its forcing function, its acceleration of joint progress in a wide range of disciplines." As a result, space R&D had contributed "significantly to the fundamental strength and viability of the United States in a world where economic and military security increasingly rest[ed] upon technology."[45]

Frutkin was at pains to emphasize that all European countries were invited to contribute to the Advanced Cooperative Project. Nor was NASA wedded to a solar probe or a Jupiter probe. If the Europeans had other ideas, they should suggest them. The important thing was to embark on a major joint scientific experiment that would also help close the technological gap. NASA officials were surprised, not to say disillusioned, by the lukewarm response to their proposal, the FRG excepted.[46] The British were particularly negative. They were not convinced that major investments in civilian space exploration were the best way to raise the general level of technology in society or to close the technological gap. The cost of the probes ($100 million–$150 million) were way beyond what European governments were willing to spend on a single space-science project. Some people in France felt that NASA's offer was driven by the need to find additional resources for space science when the expenditure on the Apollo program was peaking, rather than by a genuine wish to collaborate. On September 13, 1966, the British reported that NASA and the US government "fully understand that the Jupiter Probe is a dead duck."[47]

Faced with this disappointing response, NASA administrator James Webb went to Germany to see what could be arranged bilaterally with Minister of Scientific Research Gerhard Stoltenberg. They quickly reached an agreement for expanded scientific collaboration and for the FRG's development of a probe to the sun, which was later called Helios.[48] A few days later, Erhard was back in Washington to renegotiate the offset payments with Johnson. Part of the conversation concerned financing the probe.

September of 1966: Johnson, Erhard, and Offset

The FRG was required to "offset" with military purchases for its Bundeswehr (armed forces) the approximate US balance-of-payments deficit resulting from the deployment of American forces on German soil. From the American point of view, this arrangement both provided a market for US weapons systems and compensated for the cost to the US of maintaining a large military presence abroad. From the FRG's point of view, it secured an American commitment to help hold the front line against Soviet expansion.

The policy was not popular in Germany. For one thing, it was associated in the public's mind with a series of crashes of American-made F-104G Starfighter jets—ten in the first half of 1966 alone—that gave the impression that the US was selling unreliable and unnecessary military equipment to its ally.[49] Donald Hornig was told that it sharply reduced the domestic arms market for German industry, thereby widening the technological gap. And the sums involved were considerable. Erhard was expected to place weapons orders worth $1.35 billion by December of 1966. He was supposed to make a further $1.4 billion in offset payments by June of 1967. Meeting these obligations was particularly onerous because, to win the elections in October of 1965, Erhard had lowered taxes and increased a wide range of social benefits. This had achieved the desired political result, but it had resulted in major cutbacks in many sectors, including a 10 percent cut in defense spending.

When he met with Erhard in December of 1965, Johnson was uncompromising on the need for the FRG to meet its obligations if it expected no reduction in US troop levels. "Towering" over Erhard during their meeting, and "gesticulating and speaking in a strong strident voice," Johnson got Erhard to agree to honor the offset agreements. Domestic pressures and the damage that the war in Vietnam was doing to the US balance of payments gave US president little room to maneuver.[50] Erhard knew that he had no hope of complying. Although Erhard's government struggled valiantly to balance its budget, by September of 1966 it was estimated that the government would fall $700 million short of its obligation to place $1.35 billion of weapons orders by the end of the year and would be at least $400 million short of the additional $1.4 billion of offset payments due in June of 1967.[51]

The Erhard-Johnson meeting on September 26 and 27 of 1966 was described by McGhee as the "most critical one" yet between the two leaders. Erhard's popularity had plummeted, in part because the "Gaullists" in the Bundestag saw him as weak in dealing with Johnson over offset payments. The US ambassador, deeply concerned by the damage being done to US-FRG relations, urged Johnson to make a number of concessions to his desperate ally, including extending offset payments beyond military purchases to include "public purchases in the field of space, foreign aid and other areas."[52] National Security Adviser Walt Rostow agreed. Secretary of Defense Robert McNamara was totally opposed, insisting that all Erhard had to do was increase defense spending.[53]

The meeting in September of 1966 had uneven results. Erhard quickly conceded a major point of contention when he abandoned the demand for a multilateral force or an Atlantic Nuclear Force to satisfy Germany's nuclear aspirations. The FRG no longer sought a hardware solution to nuclear sharing, and was satisfied with participation in the Nuclear Planning Group. As for offset payments, Erhard insisted that it would "not be possible to fulfill all financial obligations under the current offset agreement to the full amount." He sought instead to stretch the repayments over an extended period. Johnson, now willing to accept non-military contributions toward offset, was opposed to that option, and became increasingly frustrated, eventually accusing Erhard of not keeping his word. The exchange closed with Erhard reluctantly agreeing to see if "more could be paid by July 1 through the purchase of computers, space equipment and items of a similar nature."[54]

The next day, Johnson made a major gesture to convince Erhard of the value of an investment in space. He and James Webb accompanied Erhard on a trip to Cape Kennedy. Two astronauts and a number of distinguished guests, including Wernher von Braun, were there to meet them. In an official address in the not-yet-completed Vehicle Assembly Building, Johnson thanked all those who had come to the US from Germany, including von Braun, for the "great efforts" they had made for the American space program. He also enumerated the many projects NASA had engaged in with European partners, and reiterated his desire to "vigorously pursue" international cooperation in space science and to provide launchers for space efforts of mutual interest.[55]

On the flight back to Washington, Webb spent an hour with Erhard. He emphasized the importance of space collaboration for closing the technological gap, and he assured Erhard that Johnson "was, in fact, offering him more than friendship and more than dollars. In fact he was offering a partnership in the development of technology that could permit Germany to increase its own capability, gain a better understanding of its own needs and opportunities for multilateral and bilateral cooperation, [and] establish a basis for leadership in the direction it felt its leadership could be effective in Western Europe." According to Webb, such a partnership could also "set a pattern of university/industry/government cooperation suited to the needs of Germany, benefiting throughout from our own experience."[56] Erhard was apparently impressed. Webb told Secretary of State Dean Rusk that Erhard's attitude seemed to have been changed by his visit to Cape Kennedy. "In fact," Webb noted, "he did say that it was impossible to learn from pictures, television, and documents the true scope and magnitude of what was being done and that he had a much better appreciation of its importance."[57] This did not help Erhard when he got home. A vote of no confidence in parliament six weeks later removed him from office.

Hubert Zimmerman dismisses the notion that the possibility of US-FRG space collaboration helped to resolve the offset crisis, noting that "the expected sums were much too low to have any significance."[58] The Johnson administration may have seen matters differently. For example, early in 1967 there was some concern in Washington that Germany would not agree to the US using some of the money available from offset "for the purchase of American wheat for India."[59] The FRG was one member of a multinational consortium that was committed to this venture, its contribution being a mere \$30 million or \$35 million. The Johnson administration seemingly attached more importance than Zimmerman allows to substituting relatively minor, in dollar terms, nonmilitary purchases for military ones to meet offset obligations. Nor should the significance of the gesture be reduced to its financial envelope. Reinhard Loosch, a jurist with extensive experience in the FRG's Ministry for Research and Technology, saw the possibility of a joint space program as something of a breakthrough as regards offset. As Loosch put it in an interview with me, it offered the option of doing something other than "just *pay,* help pay for the costs, ... something where we could get something out of it." Space collaboration with NASA also carried the prestige value that the US had hoped for and helped close the technological gap. As Loosch put it, it "gave us at least the feeling that we would then be at the forefront of technology."[60]

Conclusion

[T]here was general speculation that the principal reason for President Johnson's offer was to provide a substantial sop to compensate Chancellor Erhard for Germany not obtaining President Johnson's agreement to a share in the control of nuclear weapons.

—internal memo, UK Foreign Office, February 1966[61]

In June of 1969, more than three years after the first official contacts concerning possible joint space projects, a memorandum of understanding was signed by the US and the FRG for a spacecraft called Helios. It was the most ambitious bilateral scientific project that NASA had yet undertaken. In December of 1974 and in January of 1976, Helios 1 and Helios 2, two German spacecraft weighing about 200 kilograms each and costing about $100 million, were launched on Titan rockets from the Kennedy Space Center at Cape Canaveral into elliptical orbits around the sun. They were designed to fly closer to the sun than any previous spacecraft (approaching to within 45 million kilometers) and to provide novel scientific information about solar processes and solar-terrestrial relationships. Each probe carried ten experiments, the majority of them German.[62]

This chapter has sought to embed this project in the context of negotiations between Johnson and Erhard on offset payments, nuclear sharing, and non-proliferation. The US assumed that the desire for prestige and status was an important or even a dominant factor in Germany's nuclear policy. If that was so, technological spectaculars in space would be as effective as access to nuclear weapons in providing Germany with a sense of national pride. The opportunity for the FRG to use some of the money spent on the collaborative science project to reduce offset repayments and to close the technological gap (if only a little) was an added incentive.

To call participation in the Helios project a "disarmament" strategy, as Webb did, is surely to exaggerate its weight. But neither can it simply be dismissed as a "substantial sop" to Erhard, as the UK Foreign Office described it. The Gilpatric committee had a broad vision of what was required to curb proliferation. The Chinese nuclear test, they felt, would provoke one country after another to seek nuclear weapons, beginning with those whose security was directly threatened, and could catalyze a pro-weapons policy in the FRG. A multi-pronged attack was needed to achieve the ultimate goal of a treaty that would stabilize the international system. Technological collaboration in space science was just one of a number of incentives that collectively could contribute to stopping or at least retarding national nuclear programs, creating a climate conducive to establishing international norms that would regulate the dissemination of nuclear technology.

"Positive disarmament" sought to promote non-proliferation by technological substitution. Neither Schaetzel nor McGhee could see why any Continental ally of the US needed nuclear weapons when those countries were protected by the security guarantees provided by the US and by NATO. It was felt that their quest for independence was driven by a yearning for prestige, and that in order to make the world safer their "energies" ought to be redirected down harmless but prestigious technological paths. The appeal of this strategy in the target country depended on an already-present domestic space lobby that seized the opportunity to promote its agenda nationally—perhaps, but not necessarily, at the expense of a nuclear-weapons program. It would be foolish to place too much weight on non-proliferation by technological substitution as a disarmament policy. It is indicative, though, of the determination of the Johnson administration to curb dissemination, and of its ingenuity in using its technological assets to that end.

4

Integration and the Non-Proliferation of Ballistic Missiles: The United States, the United Kingdom, and ELDO, 1966

In February of 1966, the British government delivered an aide-mémoire to the UK's fellow member states in the European Launcher Development Organization (ELDO) officially confirming that it intended to withdraw from the organization as soon as it could.[1] ELDO had been established less than five years before, at the instigation of the British, to provide Europe with its own means of access to space. The threat of withdrawal deeply distressed Britain's Continental partners. There was consternation in Washington too. Over the next six months both European and American officials put immense pressure on Her Majesty's Government (HMG) to reconsider its decision. In May, Sir Solly Zuckerman, Chief Scientific Adviser to the Ministry of Defence, was called to Washington, where senior administrators in NASA and the Department of State made it clear that they expected the British to reverse their course. Technological incentives were offered to keep them engaged. The Cabinet changed its mind in July of 1966, and Britain remained a member of ELDO.

These events unfolded against the backdrop of major threats to the viability of the broader European integrative project. For seven months, beginning in June of 1965, President de Gaulle had withdrawn French delegates from the EEC Council of Ministers (leaving a "chaise-vide") to protest the funding of the Common Agricultural Policy and, above all, a proposal that important Council policy decisions be made by majority vote rather than by unanimity.[2] In March of 1966, in a famous telegram to President Johnson, de Gaulle announced that he would remove French forces from under NATO command.[3] This was followed by a highly publicized trip to Moscow in June. The French president attended a satellite launch from the space base in Baikonur, and endorsed a major agreement for technological cooperation in science and applications. As the historian

Marie-Pierre Rey puts it, space became the "emblematic flagship" of Franco-Soviet collaboration.[4]

De Gaulle's behavior produced a variety of reactions in the US, ranging from outrage by Under Secretary of State George Ball to feigned indifference by President Lyndon Johnson. Johnson said he wasn't going to get into a "pissing match with de Gaulle" that would only serve to boost both France and its leader.[5] The UK's assault on ELDO, however, demanded a response. If ELDO were to collapse, it would strike another blow at European integration. ELDO also had to be preserved because it was an instrument of soft power that curbed the proliferation of missile technologies. George Ball put it this way in a telegram to the American embassies in ELDO member states:

> We see advantages in multilateral frameworks for launcher development [such as ELDO]. In such a framework rocket programs tend to be more open, serve peaceful purposes, are subject to international control and absorb manpower and financial resources that might otherwise be diverted to purely national programs. National rocket programs on the other hand tend to concentrate on militarily significant solid or storable liquid fuel systems, are less open, and less responsive to international controls. Any breakup of ELDO might lead to some national programs tending in the latter direction.[6]

Herman Pollack, Acting Director of International Science and Technical Affairs in the Department of State, impressed this point on Zuckerman when they met in London in May. Others pointed out that a dissolution of ELDO might also lead to "the development of a German strategic delivery capability" similar to that which France already had.[7] And then there was the fear that "the Soviets would move into the vacuum if ELDO collapsed,"[8] providing Europe with an alternative means of access to space to that provided by the US.

George Ball's reference to types of propellants needs elaboration. The thrust of a liquid-fueled rocket is provided by mixing an ingnitable liquid (say, kerosene) with an oxidizer (say, liquid oxygen).[9] These propellants cannot be pre-loaded into the propulsion system for fear of an explosion. That did not matter for a satellite launcher. But a missile fueled with this non-storable mixture is reduced to being a first-strike weapon: because of the time required to fuel it, it might be destroyed by the enemy before it could be launched. Storable liquid and solid fuels were developed to overcome this deficiency. Ball saw an opportunity to channel European resources down a civilian technological trajectory by supporting an ELDO rocket that used non-storable liquid fuel. Interestingly enough, NASA and the Department of State adopted a similar strategy to curb missile proliferation in India and Japan at this time.[10]

Historians of European integration have paid little attention to ELDO, or to the importance that the US attached to its survival in the mid 1960s—perhaps because it was an ignominious (though instructive) "failure": it was dissolved in the early 1970s without its rocket ever having successfully put a single satellite in orbit. The political role of the organization as a technological platform for European integration also never had the historical weight of Euratom. And US support for ELDO as an instrument to curb the proliferation of ballistic missiles in Europe is overshadowed by political and scholarly pre-occupation with nuclear weapons, rather than with their delivery systems. This chapter fills these lacunae. It also exposes the limits of technological collaboration as an instrument of soft power in the sensitive domain of rocketry. Scientific collaboration is one thing. When one moves into a technological domain that touches on economic and military security, however, collaborative projects become hostage to export controls and other constraints that restrict the flow of sensitive knowledge to foreign entities. As with Euratom, so with ELDO. Both served foreign-policy and strategic objectives in Europe, embedding dual-use technological development in formal agreements and multinational structures that sought to curb proliferation. The Atomic Energy Act restricted the range of technological collaboration possible with Euratom. In the case of ELDO, US technological support for rocket development abroad was tightly controlled by national security directives. NASA and the Department of State, somewhat against their wills, were obliged to impose a number of conditions on technological collaboration that eventually undermined its appeal to their European partners. The result was a growing realization in Europe that its autonomy in a major technological field such as rocketry required that it cut the ties that had tethered it strongly to the US for more than ten years and become a more nearly equal partner—and a competitor—with the world's leading scientific and technological power.

Prehistory: The Origins of ELDO

The British were the initial driving force behind what became ELDO. They saw it as a European solution to a British problem. In 1960 the government of Prime Minister Harold Macmillan decided to cancel Blue Streak, a liquid-fueled intermediate-range ballistic missile that had been under development for about five years with American help. Two main reasons informed this decision. First, the time required to fuel a Blue Streak made it a first-strike weapon. As Stephen Twigge puts it, "in a time of crisis a choice would have to be made, to either show caution and risk being disarmed, or react immediately and risk starting a nuclear war."[11] The second reason was the cost. An internal British document circulated in February of 1960 estimated that it might cost as much as £500 million to develop, produce, and install 125 Blue Streak missiles in hardened underground silos by the mid 1960s.[12] The British government preferred to take up the United States' offer to sell it the Skybolt air-to-ground missile, then under development, which could be fired from Britain's "V-bombers." (Skybolt was later canceled, to be replaced by Polaris submarines that Kennedy offered to Macmillan when they met at Nassau in December of 1962.)

Macmillan was loath to dissolve the engineering and industrial teams that had been working on Blue Streak and to waste the money already spent on it (£67 million). He thus thought to recycle it as the first stage of a civilian satellite-launcher to be developed cooperatively with Continental partners (Satellite launchers generally combine several "stages" to enable them to escape from the downward pull of the Earth's gravitational field.) The French were the most desirable candidate, technologically speaking. In February of 1960, France had successfully tested its first fission bomb in the Sahara desert. In November of 1965, France became the third "space power" when a Diamant launch vehicle derived from the "Precious Stones" missile program put a small satellite in orbit. De Gaulle's determination to build an independent nuclear deterrent, financed by the economic expansion of *Les Trentes Glorieuses*, propelled France to the leading technological position on the Continent.

It took a good deal of effort by the British (helped by Macmillan's application to join the Common Market in July of 1961) to bring European partners on board. Everyone was concerned by the decentralized management structure that distributed the construction of the different stages of the launcher along national lines, instead of treating it as an integrated system and allocating work packages to the industries best able to do them. The French had little technological interest in the new rocket, and it took de Gaulle's personal intervention to bring them into the program. The FRG agreed to build the third stage. As Minister of Scientific Research Gerhard Stoltenberg later explained, Britain and France "had had a continuous national research and development program" in rocketry for many years. By contrast, "for the Federal Government, which had started with nothing, ELDO had provided the opportunity of entering the space field again and of reaching in time a world level with their partners."[13] Multinational, civilian, and European, ELDO enabled Germany to overcome the historic burden of von Braun's having developed the V-2 missile for the Nazi regime and to re-enter the field of rocketry.[14]

ELDO officially came into being in February of 1964. It had an unusual set of member states: Britain, five of the Six (Luxembourg excepted), and Australia (which provided launching facilities in Woomera). The financial contribution of each member state to the operating budget was fixed through hard bargaining that left Britain paying almost 39 percent of the cost. This was far in excess of what had been the norm to date in European scientific and technological organizations: usually no one member state was expected to pay more than 25 percent of the budget. The untenable exception was indicative of the dominance of political "horse trading" over technological conviction and would not outlast the Macmillan government's defeat at the polls.

The British Challenge to ELDO in the Mid 1960s

The Labour Party, led by Harold Wilson, won the British general election in October 1964. Over the next three years, the parlous state of the economy and the drain on foreign reserves led to repeated attacks on the pound, triggering crises in 1964 and 1966. The government refused to devalue the pound against the dollar, preferring instead to borrow money and to reduce public expenditures.[15] In addition, Wilson had come into office promising to cut "prestige" technological projects that made little economic sense and to buy weapons systems from the US.[16] Britain's attitude toward ELDO was directly affected by these policies.

The Labour government appointed a Committee of Officials to review all aspects of space policy immediately after coming into office. The committee presented a report within a few months. All departments agreed that the ELDO program was of "low economic priority" and of the kind that the government had pledged to cut.[17] The estimated costs of the initial ELDO launcher had already doubled from approximately £70 million to £140 million–£150 million, of which Britain's share might be as much as £60 million.[18] Even those estimates were uncertain: the French were now insisting that ELDO replace its original launcher with a simpler ELDO-A and start work at once on a new launcher, ELDO-B, that could put communications satellites into a geostationary orbit (a circular orbit, approximately 36,000 kilometers above the equator, in which a satellite appears to be stationary, and always at the same point in the sky, to an observer on the ground). The Minister of Defence was asked if he could make use of the launcher (and so also contribute to its cost). He refused. He saw no use for ELDO's rocket, and would not release any money in his budget for the UK's contribution to the organization. His parsimony was predictable. In a Cabinet meeting in July of 1960 at which recycling Blue Streak as a satellite launcher had been discussed, the Minister of Defence had offered only £5 million per year for five years toward the launcher's development costs (from a foreseen annual defense R&D budget of about £220 million–£230 million over the same five-year time span).[19] Circumstances had changed to ELDO's disadvantage in the following four years. In line with Prime Minister Wilson's quest for closer military cooperation with the United States, the British and American defense establishments had begun to work together on developing interoperable communications satellites that would be launched by American rockets.[20] Britain's military space communications systems were being integrated into an American (and NATO) system, and the Minister of Defence believed that the country had no need for independent domestic or regional space capabilities.

Political opposition from abroad to the UK's withdrawing from ELDO was to be expected. Britain's determination to wind down an organization that Prime Minister Macmillan had begged them to join only a few years earlier would surely be resented by its partners. This did not weigh heavily in Whitehall at first. Officials hoped that, since ELDO was now a "wanton waste of resources, both from our point of view, and from that of Europe as a whole," Britain's partners might (secretly) welcome the UK taking the lead in (and the opprobrium for) dissolving the organization.[21] And since Prime Minister Wilson was eager to establish closer technological links with the Continent, this might be the time to establish the principle that, as the Chief Secretary of the Treasury put it, "the Government's European policies should not be prayed in aid of cooperative projects whose sole justification was that they were co-operative."[22] Just being European and co-operative was not enough; the project also had to make economic sense.

On December 16, 1965, with the weight of the Treasury's arguments against it and only the Minister of Aviation and the Foreign Office in favor, the British Cabinet decided to withdraw from ELDO as soon as was practicable.[23] The ELDO Council, the high-level committee in which the representatives of member states negotiated policy, met four days later. Hoping to limit the political damage, the British delegate, Minister of Aviation Fred Mulley, did not show his full hand. He suggested only that, pending a more thorough evaluation of ELDO's program in March, funds should be released only for the first four months of 1966.

It is important not to lose sight of the roots of HMG's concerns. By the end of 1965, Britain had spent almost £100 million on the development of Blue Streak, first as a missile and then as the first stage of the ELDO rocket—twice what Britain's partners together had spent on ELDO to date[24]—with tangible results. The year before, in June, Blue Streak had been successfully commissioned. This meant that Britain could now develop a heavy launch vehicle of her own by combining Blue Streak with the highly successful Black Knight missile. But instead of doing so—though Mulley did not say this openly—the Treasury was financing a project whose original technological and political justification had evaporated, particularly after de Gaulle had vetoed Britain's application to join the Common Market in 1963. And this one member state—the UK —was financing ELDO to the tune of 39 percent, most of which would now to be used to develop the French and German stages of the rocket in those countries' respective national industries. It was a difficult pill to swallow.

Whatever the economic merits of the British case, in December of 1965 there was an "immediate and violent reaction from all national delegates" at the ELDO Council meeting. The German chairman described it as "the most serious blow yet struck at the Organization" and "expressed serious concern about the repercussions on European co-operation in other fields." Mulley backed down and approved the budget for the entire year 1966. The lesson was quickly learned in Whitehall: "the experience of the ELDO Council meeting dispels any illusion that the Western European Governments at heart share our view of ELDO and are only waiting for a British lead to break it up."[25]

Britain's partners let London know that they expected it to stay in an organization into which they had been dragged "squealing and kicking" in 1961.[26] They had a very different perception of the significance of the project, believing that "heavy launcher technology is something which Europe must acquire, if necessary at an uneconomic price, if advanced technological knowledge is not to become the closed preserve of the Americans and the Russians."[27] This was all the more important because there was some evidence that the US might not be willing to provide launchers for European telecommunications satellites without strings attached.

The fact that all of Britain's partners (and other industrialized countries, among them Japan) were convinced that major investments in satellite launchers were needed provided a strong argument for the Secretary of State and the Minister of Aviation to reopen the question of membership with the Cabinet early in 1966. So too did the Report of the Committee of Inquiry into the Aircraft Industry, chaired by Lord Plowden. The Plowden Report was debated in Parliament early in February of 1966. One of its main recommendations, accepted by the government, was that collaboration with Europe and development of a European aircraft industry were priorities for the UK.[28] This proposal would hardly be received with enthusiasm if the Wilson government were to withdraw from ELDO now.

These new arguments did little to sway the opposition from the Treasury.[29] The government went ahead with its aide-mémoire of February 1966 announcing Britain's intention to withdraw. It insisted that "the escalating costs of the initial programme, the absence of adequate technological advantage in the launcher to be produced, and the lack of an assured market for it" in the face of US competition would become painfully obvious in two or three years time. ELDO would produce in 1969 "a vehicle which will be obsolescent and non-competitive in cost and performance with launchers produced by the United States." Better to cut one's losses now and find more productive ways to deploy the resources available for space research and technology.[30]

The Foreign Office and the Minister of Aviation balked at having to defend this position at the next meeting of the ELDO Council in April. Labour's victory by a large majority at a general election in March of 1966 led them to hope that three new developments might sway a previously hostile Cabinet. First, they learned that Britain's partners were willing to consider a reduction in her contribution to the ELDO budget for future programs from about 39 percent to about 25 percent. Second, the French had come up with a new, more economic scheme to upgrade the ELDO rocket. This new system, called ELDO-PAS, involved adding a "high-energy perigee-apogee motor" to the system to enable the launcher to place communications satellites in a geosynchronous orbit.[31] Third, Sir Solly Zuckerman, who had been ardently opposed to Britain's staying in ELDO, was "beginning to wobble" on the decision to get out. Zuckerman reported that a British team that had recently visited Washington to discuss requirements for military communications satellites had learned that the US would "probably be unwilling to grant us sole control of any satellite they might launch on our behalf." Sir Solly added that this might also be indicative of American thinking in the commercial comsat field—a field in which he "believed that, having achieved a major monopoly position at vast expense, they would never agree to help us compete with them."[32] This was a strong argument for securing independent access to space.

The ELDO Council convened again on April 26 to debate the future of the organization. The day before, in a high-level meeting, British Ministers had agreed on a "three month stay of execution" to allow for a thorough evaluation of the costs of the perigee-apogee system (PAS) and to get more detailed information on how the financial burden might be shared.[33] This concession did not deflect the hostility of the other ELDO Ministers. They recognized that Europe had started way behind the superpowers and could not hope to compete commercially with the United States in rocketry. But that was not the prime objective in establishing ELDO, they said. They were trying, said Minister of Scientific Research Gerhard Stoltenberg, not simply to develop a competitive launcher, but to enable Europe to carry out space missions "out of its own resources, on the basis of its own knowledge and experience, at its own discretion, [and] for specifically European motives and requirements." Technological dependence "would be tantamount to conceding to the supplier country from the outset a scientific, technological and economic monopoly." The effects of this were difficult to foresee but were potentially very damaging. As the ELDO Secretariat pointed out in its submission, the heavy demands made by the government and private clients on US launch vehicles meant that there was no guarantee that NASA could launch European satellites to meet European schedules. And of course there was the risk that when commercial systems such as those for telecommunications were involved "the economic interests of the supplier country and Europe [would] cut across one another or fail … to coincide" (Stoltenberg), with negative consequences for Europe. A definitive decision on ELDO's future was postponed until June 9, 1966.[34] It was into this fluid and evolving context that, in May, NASA and the Department of State added their voices to those who were convinced that Britain should not leave ELDO.

The Positions of NASA and the Department of State

The news that Britain was seriously considering withdrawal from ELDO and the havoc that France was wreaking with the whole thrust of American policy for Europe led the US Department of State to arrange a meeting with Sir Solly Zuckerman early in May of 1966. NASA Administrator James Webb joined Herman Pollack to impress upon Sir Solly that the US had severe misgivings about the current British policy. First, it threatened to destabilize further European integration: as George Ball put it, "The United States has a direct interest in the continuation of European integration. It is the most realistic means of achieving European political unity with all that that implies for our relations with Eastern Europe and the Soviet Union … and is the precondition for a Europe able to carry its proper share of responsibility for our common defense."[35] Second, there was an increased risk that missile technology would proliferate if ELDO were to collapse. ELDO, said Pollack, "focused the attention and the resources of European nations on space projects with no immediate military application, and lessened the possibility of European efforts to develop nuclear delivery vehicles."[36] Third, a European launcher program would help close the technology gap.[37] NASA's Director of International Affairs, Arnold Frutkin, emphasized this point during visits to London and Bonn (see chapter 3) at about the same time. If the Europeans were to develop their own launch vehicles, he said, "the techniques acquired would enormously reinforce the industrial power of the West and make the European countries much more useful partners of the United States." Webb used the same argument when he met with Zuckerman in Washington.[38] European countries should not fall behind the US in technological strength, he said; rather, they should work together to "bring on underdeveloped countries and to resist communist influence and subversion."

Zuckerman had difficulty seeing technological projects in anything other than financial terms. Why, he asked, did the US want to help Europe develop something that "might oppose their own commercial interests"? Webb explained that only in this way could Europe be "a partner or client of the American space industry." It was a matter of enlightened American self-interest, Frutkin said, to strengthen Europe's technological capability even though a stronger Europe would produce competitors. "European dependence on borrowed American techniques" was "no substitute for experience directly gained and an unsatisfactory basis for any worthwhile United States–European partnership."[39] Frutkin could not forbear to point out that the Europeans would not solve their "technological backwardness, and thus contribute to the strength of the West" if they "did nothing to help themselves."[40]

These arguments caught many senior officials in London off balance. Wilson's Cabinet Secretary, Burke Trend, probably spoke for many in the Cabinet when he wrote that the US government had "perhaps somewhat surprisingly taken a decided initiative in making it clear to us that they are most anxious that we should stay in ELDO," offering technical help as an incentive. This had two drawbacks. Technological help was not what the British sought: they were against promoting a "prestige" cooperative project that seemed to have no economic rationale, all the more so given the current state of their economy. Second, the technological content of what could be shared was seriously restricted anyway by two National Security Action Memoranda, NSAM294 and NSAM338. When Zuckerman reported on his trip to Washington, he explicitly noted the constraints under which NASA and the Department of State had to operate.[41] "The Americans were not prepared to help us financially in any way," he told Wilson, though "they might give us some help in limited technological fields" suggested by the British.[42] This did not inspire much enthusiasm. Britain's main problem with ELDO was its cost, not its technological limitations.

Two Impediments to Technology Sharing in the Domain of Rocketry: NSAM294 and NSAM338

The diplomatic pressure on Zuckerman was indicative of the limited value of technological collaboration in rocketry as an instrument of soft power. As has already been mentioned, NASA and the Department of State were constrained by two National Security Action Memoranda, one issued in 1964 and one in 1965.

NSAM294 was a response to Charles de Gaulle's threat to develop a *force de frappe* that would empower him to shape France's strategy for the defense of Europe. Issued by McGeorge Bundy, the Special Assistant to the President for National Security, on April 26, 1964, it affirmed that it was the policy of the US government to "oppose the development of nuclear forces by additional states, other than those whose forces would be assigned as part of a NATO nuclear force, targeted in accordance with NATO plans and, except when supreme national interests were at stake, used only for the defense purposes of the Alliance."[43] In view of French policy, which defied all these requirements, it was in the US government's interest "not to contribute to or assist in the development of a French nuclear warhead capability or a French national strategic nuclear delivery capability." All governmental, industrial, or commercial exchanges that facilitated the proliferation of weapons of mass destruction were prohibited. Although France was specifically targeted by this directive, the Gilpatric Report of January 1965 urged that its scope be extended to "restrict United States contributions to the development of nuclear warheads or strategic nuclear delivery capabilities of *any* country (including the United Kingdom)."[44]

The difficulty posed by NSAM294 for NASA and the Department of State was that much of the technology used in launch vehicles was the same as, or bore a close resemblance to, that used in ballistic missiles. "Boosters" were a "dual-use" technology—civilian and military. Faced with the fact that a friendly country was exploring the possibility of developing its own launcher, NASA could either refuse to collaborate—and squander the leverage that US technological leadership gave the agency to shape the technological path of the program—or try to find significant zones of technological collaboration that respected the strictures of NSAM294. The latter was the preferred alternative, as expressed in Ball's distinction between solid and storable liquid fuels (good for missiles, notably those fired from submarines) and liquid non-storable fuels (preferred for launch vehicles).

This was not the only or the most onerous factor that hampered technological collaboration in the development of launchers, however. NSAM338 imposed even broader constraints. NSAM338 was issued on September 15, 1965 by McGeorge Bundy.[45] Its substance was contained in an accompanying memorandum written by General J. D. O'Connell, Special Assistant to the President for Telecommunications. Its roots lay in President Kennedy's commitment, made in 1962, to the creation of a single global satellite telecommunications system. An international organization, Intelsat, owned this system in the name of the US "for the service of all mankind."[46] The US government was represented on the Intelsat board by a private corporation, Comsat. Comsat also managed the satellite system. Interim agreements adopted in 1964 for five years specified that Comsat's share of the vote on the Intelsat board could never drop below 50 percent. This meant, as White House staffer Charles Johnson bluntly put it, "we control."[47] General O'Connell's policy, expressed in NSAM338, sought to perpetuate this situation by containing the "proliferation" of commercial comsat systems that posed a threat to the single global US-led system. To achieve that, NSAM338 restricted the sharing of technology and information that might stimulate foreign development of both communications satellites *and their launch vehicles*. Explaining a complicated policy for the benefit of the president, McGeorge Bundy noted that "the core of the proposed policy is to use our technological superiority to discourage commercial competition with Comsat and/or wasteful investment in several duplicative Free World defense-related systems."[48] Formally NSAM338 required that any government-to-government agreement on the provision of space technology, or to launch a foreign satellite, not assist in any way "the emergence of competitors who might divert traffic from the single global commercial comsat system."[49] In other words, NSAM338 sought to protect the United States' dominant commercial (and defense-related) positions by denying launch assistance of any kind to those who might cause "significant economic harm" to Intelsat.

The timing of NSAM338 was not coincidental. The first Intelsat satellite, later nicknamed Early Bird, had inaugurated direct transatlantic television, telephone, and facsimile communications just a few months before. The conquest of space was being redefined to include civilian space applications (meteorology, Earth observation, telecommunications) that had important social benefits and offered business opportunities. Determined not to be excluded from this promising domain, the French, as has already been noted, immediately called for an upgrade to the ELDO launcher (ELDO-PAS) to enable it to place a comsat payload in geosynchronous orbit. Success by ELDO in its technical mission would enable the Europeans to defy, at least in principle, the United States' attempts to retain control over a global comsat system and the associated financial, political, and cultural benefits. This was the new goal of ELDO when Britain announced its intention to withdraw. And ELDO was the European organization that NASA and the Department of State wanted to sustain by offering technological collaboration, notwithstanding the hurdles posed by NSAM294 and NSAM338.

The difficulties that lay ahead had already been sensed by Sir Solly Zuckerman in May of 1966 when he had visited Washington. Upon getting home, he reported to the Prime Minister that ELDO could not expect much help with boosters, and that any launcher that incorporated American technology had to respect the Intelsat agreements.[50] As was mentioned above, Zuckerman also came to realize that the US was reluctant to "allow the United Kingdom management and use of the United Kingdom satellites which we wish to inject into the US military satellite communications network."[51] These US conditions caused the Chief Scientific Adviser to "wobble" on the ELDO issue and to eventually come down in favor of preserving the organization—not because he sought help from the United States, but because he sought European independence.

The British Cabinet Re-evaluates Its Stance on ELDO

The Cabinet re-evaluated membership in ELDO at the end of May 1966, after Zuckerman's trip to Washington.[52] The Minister of Aviation Fred Mulley again emphasized the political arguments on the Continent against withdrawal. Now he could add that that the US government were also "anxious that [Britain] should remain in ELDO; our withdrawal might therefore forfeit US goodwill on which we were greatly dependent in many other fields."[53] Mulley failed to convince the majority of his colleagues, however. The foreign-policy argument was dismissed on the grounds that there was no advantage to be gained from "acquiescing in the wishes of other governments at our own expense."[54] The meeting again endorsed the decision that Britain should withdraw from ELDO as soon as was practicable. This conclusion was so controversial that three senior officials in the Foreign Office (Sir Con O'Neill, Sir P. H. Gore-Booth, and

Lord Chalfont) took the unusual step of formally recording their protest. The Cabinet had adopted an incoherent and arrogant policy—one that implied that "our partners are united in error and perversity," and one through which the UK was "almost gratuitously damaging our political relations with Europe at a time when our great concern should be to nurture them" in anticipation of a new application to join the Common Market.[55] The objections made by the three fell on deaf ears.

This drawn-out and contentious saga turned to farce within a fortnight. On June 3, 1966, the London *Times* published the Cabinet's decision, which had been leaked to the newspaper in anticipation of the formal announcement at the ELDO Council meeting at Ministerial level a week later. Whitehall was inundated with telegrams from the UK's partners regretting the move. Dean Rusk "expressed the profound disquiet of the United States Administration" at the technological gap in the space field that was opening up between the two sides of the Atlantic.[56] The Society of British Aerospace Companies submitted a strong statement that the government's position, "as well as having a disastrous effect on British space activities[,] ... could demolish at one blow all faith in Britain as a suitable partner in collaborative aerospace programmes and could jeopardise the plan, outlined in the Plowden report, for the future of the British Aerospace Industry."[57]

There is no evidence that these negative reactions had any effect. It was a new assessment by the Legal Experts of the Crown, who had been called upon to calculate the costs of British withdrawal that changed the situation completely. At the suggestion of ELDO's legal advisers, for the first time they did not limit themselves to the ELDO Convention and Financial Protocol but probed back to the "travaux préparatoires" (preparatory discussions) that the future member states had undertaken in 1961. The lawyers found that Britain had imposed strict conditions to stop its partners from reneging on their engagement to the initial ELDO program—conditions that came back to haunt Her Majesty's Government. There was a serious risk that if Britain were now to default on the program for which it had fought, as the Cabinet recommended, and the issue were to go to arbitration, Britain would lose. The cost of losing would range from £30 million to £50 million, depending on the penalties incurred if the other member states decided that they were unable to continue the initial program without Britain. By contrast, the cost of staying in with a reduction in Britain 's contribution to 25 percent was about £42 million.[58] In the best possible scenario, Britain would therefore "lose" only £12 million if it stayed in ELDO; withdrawing might be more expensive, with immense negative political consequences.

The die was now cast. The British Cabinet secured a postponement of the next ELDO Ministerial meeting to the afternoon of June 9. It met at 10 o'clock that morning and authorized Fred Mulley, who had to leave by 11:30 in order to get to Paris in time, to announce that Britain had decided to remain in ELDO under certain conditions. No mention was to be made of the new legal arguments that had swayed the government.[59] Mulley made it to Paris in time and achieved a reduction of the UK's contribution to ELDO to 27 percent.[60] The balance of the cost was taken up by Britain's partners (Australia excepted) proportional to their gross national products.[61] ELDO was put fully back on its rails at the Ministerial Conference on July 7–8.[62]

Three weeks after the aforementioned meeting, on July 29, 1966, Walt Rostow, President Johnson's Special Assistant for National Security Affairs, signed NSAM354, authorizing US co-operation with ELDO.[63] That memorandum noted that the Department of State urgently sought a clarification of the administration's policy on ELDO. It also affirmed that it was "in the US interest to encourage the continued development of ELDO through US cooperation," the modalities of which had to be defined by interdepartmental agreement. This would be conveyed to the European Organization "in the event that ELDO desired cooperation with the US."

In advocating this path of action, the Department of State believed that the British Cabinet's decision not to withdraw from ELDO was "no doubt influenced by the severe European reactions to the possibility of British withdrawal and to the expressions of US interest in the continuation of ELDO."[64] US officials were apparently not aware that purely financial arguments had tipped the balance against withdrawal. They were also not aware that, although the promise of US technological support was often mentioned as one argument for staying in ELDO, Solly Zuckerman and others also emphasized that ELDO ought to be as independent of the US as was possible. Britain stayed in ELDO primarily to avoid paying the financial penalties for leaving it, and only secondarily to remain on good terms with its Continental partners and with the United States.

NASA's Offer to ELDO

The general outlines of the United States' policy toward ELDO were defined by the Working Group on Expanded International Cooperation in Space, which was introduced in chapter 3. Its brief was to propose how best to foster collaboration as the space programs of advanced countries matured beyond scientific exploration and gained economic and political significance. "If our expectations for expanding space cooperation are to be realistic," the working group's report to the National Aeronautics and Space Council read, "we must be prepared to accept increasing foreign participation in the technological and applications aspects of our programs."[65]

ELDO's technological needs obliged a revision of this kind. It was a response to the growing interest in foreign countries in space applications and launcher development. The working group defined a list of activities that accompanied Rostow's NSAM354 and that specified the kind of support that could be offered as an incentive to Europeans to continue with the organization.[66] That support was, of course, conditional upon the constraints, imposed by NSAMs 294 and 338, that restricted the proliferation of communications satellites and guided missiles. Richard Barnes, who worked closely with Arnold Frutkin in NASA's International Affairs Division, argued for a more consistent application of NSAM294. He particularly wanted cryogenic liquid propulsion systems to be exempt from export controls to Europe, since everyone knew that they weren't used in French missiles. He also wanted US export policy to reflect the advanced state of European rocketry (particularly in guidance and control). In general, Barnes wanted technology sharing to be denied only for a few critical items that were clearly intended for use in a national missile program and weren't available outside the US.[67]

NSAM338 was not as easily dealt with. It required government-to-government agreements to guarantee that ELDO launch vehicles, components, and technology that benefited from US technical support would not be "used by foreign countries for development of their own communications satellite capabilities except (a) in connection with the Intelsat system or (b) to permit participation in the US national Defense communication satellite system."[68] Respecting that framework, NASA was willing to train ELDO nominees in various aspects of project management, a field in which Europe was weak.[69] NASA also was willing to facilitate export licenses for ELDO so as to give it priority over European national launch-vehicle programs. (This policy was similar to preferring bilateral agreements with Euratom to bilateral agreements with its individual member states so as to incentivize national commitments to the multinational organization.) A technical office would be established inside NASA to expedite the sharing of launch-vehicle technology from the Atlas rocket, and to procure unclassified hardware such as the strapped-down guidance package used on the Scout vehicle. In the long term, NASA would assist ELDO when it began to develop high-energy cryogenic stages (i.e., using a non-storable liquid propellant), would share information on the Atlas-Centaur program, and would bring ELDO personnel "into intimate touch with the problems of systems design, integration, and program management of high-energy upper stages such as the Centaur."[70]

This list of possible areas of assistance was meant to make ELDO more attractive to its partner governments. The Department of State pleaded for a more flexible regime. Even in the most liberal interpretation, it asserted, the requirements of NSAM338 were "cumbersome, torturous and irritating."[71] "Removing restrictions on the release to foreign countries of communication satellite technology now would not appreciably affect attainment of United States objectives," one paper insisted.[72] On the contrary, it would open markets, it would improve the engagement of European governments in Intelsat, and it would strengthen the hand of the United States when the definitive Intelsat agreements were negotiated, beginning in 1969. NSAM338 should be rescinded or modified; as it stood, it was more likely to encourage rather than curb the proliferation of comsat capabilities. General O'Connell was asked to revisit NSAM338 to see if the strictures it imposed could be relaxed, as they had "become a political irritant to a number of European countries." In fact they were imposing limits on the United States' ability to cooperate internationally in space "to a degree no longer justifiable."[73] O'Connell's response was unambiguous: An extensive review by the responsible agencies had concluded that "the policy expressed in NSAM338 [was] not a deterrent to cooperation, nor [was] it contributing to the technology gap between the US and friendly countries."

By coincidence, on the same day that this was reported to President Johnson—July 10, 1967—a working group set up by ELDO to consider the help being offered by NASA delivered its report to the ELDO

Council. It insisted that any technological cooperation with the US "must not involve any restriction on the use by Europe of launchers incorporating jointly developed or American developed hardware."[74] This included the broad requirement that such technology "not do significant economic harm to Intelsat."[75] Europeans would not accept American technology if they felt that the invocation of Intelsat clauses would prohibit them from using their own launchers to put their comsats in orbit. François Xavier Ortoli, Gaullist and French Minister of Industrial and Scientific Development, was quite blunt about what he saw as the unacceptable reach of American power in an exchange with his German counterpart in 1971. Speaking of the United States' restrictive policy for the launching of telecommunications satellites, Ortoli fumed:

> ... American policy in this domain is absolutely clear: it aims, on the contrary, to establish and to perpetuate its domination so as to give the United States an instrument of influence and of political action on the rest of the world at the same time as American industry benefits from a quasi-monopoly or, at least, a dominant position concerning all commercially viable applications in a field that is of the utmost importance for technological development.[76]

Technological collaboration on launchers capable of orbiting geostationary satellites was hemmed in by conditions that made it incompatible with Europe's reason for embarking on ELDO in the first place: to secure autonomous access to space.

Concluding Remarks: The Limits of Soft Power

The final report of the inter-agency working group seeking ways for the United States to expand international cooperation in space highlighted the dilemmas surrounding technological collaboration in rocketry. Space technology, it emphasized, had "unusual if not unique possibilities for arms control measures and for serving our arms control objectives by channeling foreign technological efforts into non-military activities, often with multinational participation."[77] This was non-proliferation by positive disarmament, using technological sharing as an instrument of soft power. NSAM294 constrained technological collaboration by specifying that any knowledge that enhanced France's national missile program—or indeed any foreign missile program—not be shared. NSAM338 constrained collaboration even more by seeking to ensure through bilateral government agreements that any technological assistance not "encourage the proliferation of communication satellite systems" outside the Intelsat framework. That policy impeded collaboration with ELDO even though the organization was building launch vehicles, not satellites. ELDO could not guarantee that it wouldn't be asked by one of its member states to launch communications satellites that would compete with the Intelsat system.[78] Positive disarmament was an ingenious arms-control strategy, but it came up against strict limits when it promoted the sharing of technologies that could undermine the United States' global leadership and commercial interests in communications satellites. Nor did NASA or the Department of State succeed in having those limits revised. Once again, whatever potential technological collaboration had as an instrument of soft power was squandered by irresolvable disagreements within the US administration.

Closing the transatlantic technological gap was a major theme in the negotiations over space collaboration. For Frutkin and Webb, technological collaboration was a matter of enlightened self-interest: it would foster relationships between American and European aerospace companies. Trevanion Nesbitt, speaking in Munich on behalf of the Department of State, looked forward to the day when a "combined European effort through ELDO might lead eventually to a new and useful contribution to the Free World inventory of launch vehicles as well as to an independent European launch capability"—independent but "consistent with the Intelsat arrangements."[79] This was not the "independence" the Europeans wanted. Responding to the UK's aide-mémoire of February 1966, the Dutch government insisted that it sought "equitable Atlantic cooperation"—a "two-way traffic in know-how," as another delegation put it. The Belgian Minister of Foreign Affairs, Pierre Harmel, hammered the point home. The European science ministers, meeting in January of 1966 under the auspices of the OECD, were, Harmel said, convinced that a "dialogue with the United States would only again become fully fruitful when Europe had reduced its lag and so strengthened its partner's willingness to cooperate." "How," he asked rhetorically, "can we prove to the Americans that we should be worthwhile partners if by ourselves we have never achieved anything of significance?"[80] Reciprocity was the sine qua non for sound technological collaboration. To close the technological gap, it would first be necessary to advance launch-vehicle technology inside Europe enough so that European countries could negotiate with the United States from a position of strength. In so doing, they would dilute the soft power inherent in the United States' technological leadership and deal with NASA as partners rather than pleaders.

The United States' offer of technological assistance was a misguided effort to change British policy. The American authorities placed their faith in technological leadership to achieve political objectives. They did not realize that HMG had stayed in ELDO not to secure US technological assistance but to avoid the onerous financial penalty that might ensue upon withdrawal. Indeed, Sir Solly Zuckerman, like some of his more fiery Continental counterparts, was inclined to see technological assistance as a trap—a mode of control that would unduly curb HMG's freedom of action in launching both military and commercial telecommunications satellites.

For almost twenty years after World War II, the West Europeans had snuggled within what the historian N. Piers Ludlow calls "a protective Atlantic cocoon"—an international political, economic and security system that took American leadership for granted. From the mid 1960s on, by contrast, "the overarching system began to show signs of strain," beginning with "the way in which détente, growing Western European confidence, and the United States' ever-greater involvement in Southeast Asia started to erode European willingness to accept US foreign-policy leadership."[81] Britain too could no longer be counted on to be Washington's "loyal lieutenant" (Ludlow). Zuckerman was less confrontational than François-Xavier Ortoli, but he too was aware of the role played by technological leadership in consolidating American hegemony in Europe—and his opposition to staying in ELDO "wobbled." In the next chapter we will again find Zuckerman facing, and resisting, US attempts to steer British technological choices, this time regarding the use of gas centrifuges to enrich uranium.

5

Classification, Collaboration, and Competition: US-UK Relationships in Gas-Centrifuge Uranium Enrichment in the 1960s

Controlling the production and circulation of fissile material was a linchpin of the United States' non-proliferation strategy for the first twenty years after World War II. At first the aim was to secure a monopoly of nuclear raw materials. Even before the war was over, General Leslie Groves, the director of the Manhattan Project, set up, along with the British, a materials consortium, the Combined Development Trust (CDT), to secure access to the world's deposits of uranium and thorium ores.[1] In December of 1945, Groves estimated that the CDT controlled about 97 percent of the high-grade uranium ore from existing producers. Belgium joined the group to ensure that the rich deposits found in the Katanga Province of the Belgian Congo would also be kept away from the Soviet Union.

The curb on proliferation by this near-monopoly of access to ore was complemented by the massive investments needed to enrich it. The historian R. Scott Kemp reminds us that the gas-diffusion plant at Oak Ridge that produced fissile material for the atomic bomb "employed at its peak some 12,000 people, enclosed forty-four acres under a single roof, and by 1945 consumed nearly three times the electricity of the highly industrialized city of Detroit."[2] The day the bomb was dropped on Hiroshima, President Harry Truman characterized the Manhattan Project as a "marvel" combining "the achievement of scientific brains" with "the capacity of industry," adding that it was "doubtful if such another combination could be got together in the world."[3] The Soviet bomb of 1949 shattered this illusion and disabused General Groves of an "unassailable faith in American technological 'know-how.'"[4] Security now lay in competing for technological pre-eminence protected by classification, and Truman moved quickly to authorize the development of the hydrogen bomb.

The twin pillars of techno-industrial superiority and tight controls on knowledge flows provided by the McMahon Act of 1946 served the cause of non-proliferation in the 1950s and the 1960s. Britain and France—major industrialized countries with extensive wartime experience in nuclear research—enriched uranium for an independent nuclear deterrent, but they posed no threat to the Western alliance. Non-weapons states with aspiring nuclear programs preferred to acquire their special nuclear material from the US Atomic Energy Commission, accepting the restrictions on use imposed in bilateral agreements and safeguard provisions. In short, the cost and technological complexity of constructing and operating diffusion plants, a refusal to share technology with even close allies, and the imposition of safeguards against the diversion of fissile material from civilian to military programs were the defining features of the AEC's non-proliferation strategy in the early Cold War.

In 1960, a new enrichment technology, the gas centrifuge, emerged as an attractive alternative to gas diffusion. It defied the logic of the above-mentioned policy. A centrifuge system cost less to build than a gas-diffusion system, had lower operating costs, and was technologically less complex. It offered a substantial reduction in the cost of a kilogram of enriched uranium. A centrifuge plant was also far easier to conceal than the giant gas-diffusion plants that had been inherited from the Manhattan Project and thus made it easier for defiant states to produce enriched uranium clandestinely. In short, the gas centrifuge dramatically increased the number of nations that could seriously consider having their own uranium-enrichment programs.

Gas-centrifuge enrichment both challenged the United States' monopoly of the supply of enriched uranium on the world market and posed major proliferation risks.[5] The US was not ahead of other countries in the exploitation of the technique in 1960. The UK, the FRG, and the Netherlands were all developing gas-centrifuge enrichment, and the USSR already had a plant in operation. The AEC was convinced that the security of the United States required getting "back into a position of technological leadership."[6] The US wouldn't be able to stop determined competitors from embarking on centrifuge enrichment. The best it could hope for was to slow down their efforts until an international agreement limiting the proliferation of nuclear weapons was adopted.

This chapter will discuss the legal and technological policies considered and implemented by the AEC in consultation with other branches of the US government to monitor and eventually to retard the development of centrifuge technology in Western Europe in the 1960s. Among the policies were classification of existing research, collaboration to keep an eye on what others were doing, and forcing open the black box of the UK Atomic Energy Authority's prototype production centrifuge in order to assess its performance. The bruising debate that ensued between American and British engineers and officials throws light on the leverage provided by technological collaboration from a position of strength—that is, when a dominant state's technological soft power is "hardened" by being embedded in a mutually accepted legal framework and in a network of technological dependencies that severely restrict the junior partners' room to maneuver.

The Early History of Gas-Centrifuge Enrichment

The idea of using gas centrifuges spinning at very high velocities to separate fissile U^{235} from the far more abundant U^{238} had already been looked into before World War II.[7] The principle was simple: As a bowl containing the isotope mixture spun rapidly on its own axis, the concentration of the heavier isotope would increase from the center of the cylindrical rotor to the wall. By extracting the slightly enriched mixture at an appropriate point on the radius and recycling it many times through centrifuges connected in series (a cascade), one could significantly improve the concentration of U^{235} in the mix. The path from principle to practice was arduous, however. Jesse Beams at the University of Virginia got the first encouraging results with a prototype centrifuge in 1934. The technique was tried during the Manhattan Project, but was rejected in favor of gas diffusion. Classified research began again in the mid 1950s in response to improvements of Beams' design. The work was guided by theoretical analyses which showed that the separation power of the centrifuge was proportional to its length and to the fourth power (in practice the square) of the rotational velocity of the wall. Beams extended the length of his rotors and devised ways of overcoming vibrational instabilities at high speeds. A parallel unclassified project headed by the Austrian physicist

Gernot Zippe was under way from August 1958 to June 1960 at the University of Virginia.

Gernot Zippe was one of three former German prisoners of war who had worked on centrifuges in the Soviet Union with Soviet engineers.[8] Together they had advanced the technology to the point where, by January of 1958, the Soviets were able to build a pilot plant with 2,435 centrifuges that could produce enough highly enriched uranium for one Soviet implosion-type bomb every two years. By that time the Germans had been repatriated. Zippe was interrogated by officials from the AEC, who were impressed by his novel designs for short-bowl centrifuges. He was given a contract to do unclassified work at the University of Virginia, though he was not allowed access to the classified project led by Beams. By June of 1960, he had reproduced a slightly smaller version of the Soviet centrifuge. Experts from private firms and the AEC "marveled at the device's simplicity." Zippe left the University of Virginia to work for Degussa in West Germany. In November of 1960 the AEC awarded the Union Carbide Nuclear Company a contract to demonstrate a centrifuge spinning at 450 meters per second and to test the functioning of several such centrifuges connected in a cascade.

The AEC also established that groups in Germany (at the firms AEG and Degussa, at the University of Bonn and the University of Kiel, and at the Max Planck Institute at Aachen) had "the most extensive and most complete gas centrifuge program in the world at this time."[9] Jacob Kistemaker's mass-spectrography lab in the Netherlands was also actively engaged in research. The British, who had stopped studying centrifuges in 1948, were interested again after seeing how much progress had been made at the University of Virginia.[10] None of this work was classified.

The Push for Classification

Toward the end of 1959, the AEC asked the General Electric Company to evaluate the promise of the gas centrifuge. The report's findings, widely circulated inside the AEC in April of 1960, revealed that, thanks to major improvements made over the past two to three years at home and abroad, the US would be able to build a gas-centrifuge plant capable of producing enriched uranium at a price "competitive with the present published AEC price list" if it were to invest in a further three-year development program.[11] The AEC also asked labs at Hanford and Oak Ridge to study how the gas centrifuge might affect *N*th-country proliferation relative to the reactor-based plutonium route to a bomb. The study concluded that, overall, the "gas centrifuge route is the more attractive and perhaps easier route for an *N*th country."[12] A report by Union Carbide fleshed that out, dividing candidates for weapons development into three country groups. Countries in the group X (e.g., West Germany, Sweden, and Japan) had a relatively high degree of technical competence and industrial capacity. Countries in group Y had some indigenous capacity but would probably have to import some technology and equipment. Countries in group Z (e.g., Egypt and Pakistan) had to rely heavily on external support to develop a bomb. It was estimated that the time required to produce its first atomic weapon would be 5 years for X-group countries, 6½ years for Y-group countries, and 8 years for Z-group countries.[13] A staff report to the General Manager of the AEC recommended that a development program costing about $6 million over three years should be launched at once to "place the US back into a position of technological leadership in this area of isotope separation, a subject of considerable military and commercial concern."[14] It was noted that the FRG and the Netherlands were also "vigorously pursuing studies in this field," their capacity now "equal to or excel[ling] our own state of the art in the field."[15] The report also recommended that the AEC explore at once "agreements for classification, export controls and collaboration … with the United Kingdom, Germany, and the Netherlands" to prevent gas-centrifuge information and equipment from becoming available to other countries that could use the process to produce weapons-grade materials.[16]

Later in 1960 the JCAE was briefed on the state of play. The proliferation dangers were again emphasized. Data were produced to suggest that a centrifuge plant costing at most $150 million could produce 3,000 kilograms of enriched uranium—the same amount that would be produced by the French gas-diffusion plant at an estimated cost of $750 million. Rudimentary technological capabilities were required. George Kolstad, chief of an AEC research division, said that the requisite technological capabilities were "probably equivalent to [those required for] building and operating a plant employing about a thousand people to manufacture washing machines, refrigerators, automobiles and the like." There might be "a slightly higher advanced technology involved," but centrifuge production was within the industrial capabilities of many nations. "At first blush," said Senator Albert Gore of Tennessee, this "certainly gives me an uneasy, uncertain, almost frightening feeling for the world when you consider the irresponsible people who have come to power through coups [e.g. in Cuba] and otherwise."[17] Did the European countries recognize "the desirability of achieving uniform classification policies and practices within the four countries having gas centrifuge research and development programs"? Yes, though there were "certain psychological-political difficulties to be overcome," as AEC Chairman John McCone put it.[18]

William Burr has written extensively about the "psychological-political" difficulties that the Atomic Energy Commission and the Department of State confronted in 1960 as they sought to secure an agreement with groups in the FRG and the Netherlands to classify research and to impose export controls.[19] Private industry was eager to exploit the commercial potential of the new technology, but saw classification and export controls as a "dead hand of bureaucracy" that would stifle its efforts.[20] As McCone put it, "we can keep our industry in control if others do but we can't if others don't."[21] The Dutch seemed to think that they could easily classify centrifuge research, though their industry would probably have to be compensated for potential loss of revenue. The FRG posed problems of a different order. Quite a bit of the research in Germany was done in university labs and could not be classified: Students refused to work under such a regime, though perhaps they could be encouraged to self-classify their research findings. A more serious problem with international ramifications was that if Germany were to declare publicly that it was classifying centrifuge work many would assume that it was embarking on a military program. Perhaps Euratom could be engaged to assure other countries that the centrifuge arrangements were not a cover for a military program. The problem with this approach, in the Department of State's eyes, was that the information would have to be shared with France too, something that was against AEC policy on security grounds. It seemed that the best that one could hope for was safeguards applied by the International Atomic Energy Agency along with informal commercial restrictions.[22]

These thorny problems were resolved, somewhat to the surprise of the United States, when the West German government agreed to impose classification on centrifuge work in July of 1960. The Germans and the Dutch asked the US AEC to provide them with a classification guide, which was quickly sent. The guide distinguished unclassified from confidential and secret (i.e., classified) research. It classified information on cascades, on separation efficiencies, on the design of bearings, and on any centrifuges that represented "an advance in the state of the art beyond that represented by or described in the published literature as of the effective date of this guide (August, 1, 1960)."[23] The one remaining point of disagreement was whether the four participants could file classified patents in other countries (the Americans wanted filings restricted to countries with which it had agreements to exchange classified data; the Dutch wanted to file in any country that had adequate security arrangements).[24] This was left in abeyance. On October 4, 1960, the FRG imposed state secrecy classification on centrifuge developments that were sensitive (i.e., that might facilitate proliferation of nuclear weapons), taking care to reassert its own 1954 commitment to abjure nuclear weapons.[25]

The US AEC came to a "gentlemen's agreement" with the FRG and the Netherlands in which they agreed not to disseminate their results on the centrifuge process for the time being and to consult with each other regularly. The British, however, sought closer collaboration with the AEC as foreseen in the 1955 US-UK Agreement for Co-operation on the Civil Uses of Atomic Energy.[26] It was on that basis that, from late 1960 to mid 1965, British nuclear engineers exchanged classified gas-centrifuge information with their American colleagues.

In 1964 the AEC received an updated report on the Nth-country proliferation risks of gas-centrifuge enrichment. Candidate countries were again divided into three groups, X, Y, and Z, but now it was assumed that classification excluded them from knowledge of the centrifuge process. It was estimated that knowledge of US developments could reduce the time needed to produce weapons-grade uranium (90 percent U^{235}) from eight years to five for countries in group X, from twelve years to seven for countries in group Y, and to about nine years for countries in group Z.[27] It was imperative, therefore, to maintain "current restrictions on dissemination of information on centrifuge technology." AEC Chairman Glenn Seaborg was advised by William Foster, the Director of the Arms Control and Disarmament Agency, to "resist all pressures to release controls on the dissemination of gas centrifuge technology" for as long as he could in the interests of security. Seaborg imposed new restrictions on industry that were further tightened up in 1967. No new firms were allowed to enter the enrichment field; Seaborg feared that their promotional activities might stimulate "further interest abroad in the gas centrifuge process, thus leading to a proliferation of capabilities to enrich uranium for weapons."[28] The AEC would continue to develop the technology with firms under its supervision to ensure that the US would retain "world leadership in uranium enrichment," but it would "retain strict classification controls on US gas centrifuge process information" and would "continue to support the present understandings with foreign governments whereby similar controls are now applied to their information."[29]

Secretary of Defense Robert McNamara endorsed a "delay policy." He was aware, he wrote Seaborg, that the "controls which can be placed on technology can only retard, not prevent, its growth and diffusion." Nonetheless, three kinds of steps could be taken: (1) dampen the incentive of (advanced) group X and Y countries to develop their own gas-centrifuge technology by letting them know that "enriched uranium w[ould] be available from this country on attractive terms," (2) "inhibit assistance to less developed countries" (effectively group Z) by restrictive classification policies and export controls imposed by all those who had mastered the technique, and (3) take measures to "support US gas centrifuge technology at a high level so that the US can stay abreast, or ahead, of developments in other countries," protecting this advanced technology with "stringent security controls."[30]

By 1965 the team at Oak Ridge had tested nine centrifuge models built from different materials (aluminum, titanium, aluminum-fiberglass). They had begun with a Zippe model 3 inches in diameter and 305 millimeters in length and spinning at 350 meters per second. By July of 1965 they had a 10-inch-diameter bowl spinning at 550 meters per second.[31] The arrangement with the UK was terminated by the AEC on June 30, 1965 "by a declaration that information now being generated in the US development program represented production technology and was not transmissible under the bilateral."[32] The UK Atomic Energy Authority continued research on its own and by 1967 was reasonably satisfied that it could produce enriched uranium using centrifuges more economically than with gas diffusion.

The UK's Early Moves Toward a Tripartite Agreement

By the summer of 1967, the UK Atomic Energy Authority was faced with a choice between increasing the capacity of its gas-diffusion plants at Capenhurst and investing in a gas-centrifuge plant that might reduce the cost of enriched uranium by as much as 30 percent. The US AEC, whose R&D program was probably more advanced that that in the UK, had expressed skepticism about the economic possibilities of the centrifuge. Their view wasn't taken seriously in Britain, where officials suspected that it was meant to "conceal … concern about proliferation of weapons potential and about competition from other countries with the large existing US diffusion plants for the supply of enriched uranium for peaceful purposes."[33] Britain would have to make up its own mind. To that end, the UKAEA would first have to build a pilot plant, operating by the end of 1967, that would connect individual centrifuge units into modules (a cascade) that could produce useful quantities of enriched uranium. In the meantime, the UKAEA deemed it crucial to conceal its intentions as much as was possible. Sir John Hill, the chairman of the UKAEA, told the government that the most important secret to keep, for both proliferation and commercial reasons, was that a country that already had a gas-diffusion facility was considering developing a gas-centrifuge plant—information that would signal to competitors that the British thought the gas-centrifuge technique to be viable, and would undermine what lead the UK had over its Continental rivals.[34]

The UKAEA's prudence sat uneasily alongside Prime Minister Harold Wilson's determination, driven by ongoing economic crises, to draw closer to Europe. Early in May of 1967, the Wilson government made another application to join the European Economic Community. That was followed by the Prime Minister's proposals for a European Technological Community in his annual speech at the Guildhall on November 13. Wilson emphasized that this new "community" was not a substitute for membership of the EEC, but rather "a catalyst to a deeper and closer economic integration."[35] In the view of the historian Helen Parr, it was more than that; it was a "lure to get Britain into the EEC and a stick with which to castigate French exclusionary policies." Wilson would use the UK's technological strengths in aircraft engines, computers, and civilian nuclear power to establish bilateral or multilateral agreements with the

EEC that would strengthen its economy and enable it to compete more favorably with the United States. French policies, "by repelling Britain, subordinated Europe to superior American power."[36]

President de Gaulle again vetoed Britain's application to join the Common Market at a press conference on November 27, 1967. His refusal was expected, but, in contrast with the veto in January of 1963, it was not accepted uncritically by the five other members of the EEC which had already endorsed the principle of EEC enlargement. Meeting on December 19, the Council of Ministers of the Six issued a communiqué that made it clear that only France opposed the extension of the EEC. The next day, the British Cabinet agreed that, notwithstanding the setback, Britain maintained "the longer-term objective of full membership as a valid and credible policy in the eyes of public opinion here and in Europe."[37] In February of 1968, in line with that objective, Sir Solly Zuckerman was asked by Prime Minister Harold Wilson to conduct "an enquiry into the means by which we might best exploit our civil nuclear capability in relation to our attempt to achieve membership of the EEC."[38]

The Dutch and the Germans were obvious partners; they could inject new ideas and insights into a collaborative gas-centrifuge enrichment venture, rather than simply taking advantage of what the British had already done. For some time they had been wary of the constraints imposed on their research by the informal agreement with the AEC. Seaborg's determination in 1964 to limit domestic industry's access to classified knowledge, along with a liberal fuel-pricing policy, led Bonn to believe that there had been a breakthrough in centrifuge research that would make it highly competitive with gas diffusion. To secure autonomy from US supplies, the Germans decided to try to launch a joint venture with the Dutch, a project that petered out late in 1966.[39] A year later the climate had changed. In February of 1968, researchers at Royal Dutch Shell and at Werkspoor Inc. announced at a meeting of Foratom (an association of mostly West European nuclear industries) that they had developed a faster, cheaper method than gas diffusion for enriching uranium. They had gone public, it seems, to quash pressure by the French on the Dutch to participate in the expansion of their gas-diffusion plant at Pierrelatte that had been built to meet the needs of the members of Euratom. Newspapers reported that the two firms planned to invest the equivalent of $80 million in a prototype centrifuge plant near Amsterdam, creating "something approaching consternation" in Europe, according to the *Washington Post*. The Germans were also openly speaking of their centrifuge work and of a new nozzle process that was "based on compressing a gaseous compound of uranium, and forcing it through a narrow curved channel," collecting the lighter U^{235} isotope from the inner side of the curve.[40] The FRG already had a plant with a dozen stages, and saw no reason why it couldn't be scaled up.

The UKAEA's original reaction to these announcements was to maintain the centrifuge "secret." It also tried to discourage the Dutch and the Germans by claiming that work on centrifuges done by the UKAEA had produced disappointing results. A similar tactic had been used by Glenn Seaborg to put the UK "off the scent" of centrifuge enrichment. Minister of Technology Tony Benn pointed out to Prime Minister Wilson early in March of 1968 that the British were doing to the Dutch and the Germans what the Americans had done to them.[41] Britain's lead, such as it was, was a wasting asset. The UKAEA should abandon its "false position," come clean about its optimism about the commercial prospects for gas-centrifuge enrichment, and begin discussing a joint project as soon as that was possible.

Resentment of the French and suspicion about their intentions provided an added incentive. A report prepared for the Prime Minister in February of 1968 suggested that "something [was] going on between the European powers in the nuclear field of which we are not fully informed—something in the nature of ganging-up against us which derives basically from the French desire to establish a nuclear pre-eminence in Europe and to make her partners in the EEC particularly dependent on her in this context." The proximate cause of the suspicion was the Franco-Belgian decision to buy an American reactor rather than a British one. Zuckerman also thought that the French were pursuing leadership in nuclear matters "partly, if not primarily, in order to bind the Germans more closely to them." These two initiatives occurred simultaneously with de Gaulle's increased determination to cut Europe free, under French leadership, from the binary logic of the Cold War and from a putatively narrow concept of détente. His new military doctrine, announced by General Ailleret in December of 1967, engaged the country in an expanded "all around" (*tout azimuts*) concept of deterrence that would allow the French to intervene anywhere, and not just against a specific enemy.[42] It was received in

NATO and by the Atlantic Alliance as something between a provocation and a threat. De Gaulle was becoming increasingly isolated, his international prestige undermined even more after the tumult of May 1968. Here was an opening for Britain. As Tony Benn put it, the French government was "as hostile to our technological approaches and proposed links with the Six as [it was to Britain's] political approach to membership of the community."[43] At the same time, France was having difficulty with many major technological projects— the second stage of the ELDO rocket was not working as had been hoped, and their gas-diffusion system was wholly uneconomical and uncompetitive with the one at Capenhurst, let alone with centrifuges. By cooperating with other members of the EEC at France's expense, Britain could "add a technological isolation to their present and obviously growing political isolation." A joint uranium-enrichment project using gas centrifuges would breathe life into an emerging technological relationship, would "isolate the French with their high-cost, low efficiency nuclear technology," and would "secure for [Britain] a commanding position in the nuclear field in the Six."[44]

A number of obstacles stood in the way of the UKAEA's and the British government's seriously embarking on the aforementioned scheme. The inclusion of Germany might antagonize the Soviet Union. As a French Minister remarked to Tony Benn, collaboration with the FRG was difficult—"until our Russian friends accept Germany into the community of nations, we do not want any suspicion that they are getting access to nuclear weapons technology."[45] The British would also have to begin treating Germans as equal partners. They had previously asked them to participate financially in the extension of the gas-diffusion plants at Capenhurst in return for a guaranteed supply of enriched uranium and some management control—but no access to technology. It was not surprising that the UKAEA were still waiting for a response a year later. Christopher Audland, the Head of the Science and Technology Office in the Foreign and Commonwealth Office, told the Cabinet that they should realize that a new generation of people who felt no personal responsibility for Germany's past behavior were gaining influence in the FRG and were not willing to be dealt with disparagingly. The British "could wean the Germans away from tending their relations with the French, and establish a stronger claim to their sympathy [if] they were more understanding of their aspirations than are the French." The Germans, for example, had a quite different approach to ELDO. The threat by the British to withdraw from that organization in spring 1966 had been strongly criticized by Gerhard Stoltenberg, the Federal Republic's Minister of Scientific Research. It was Stoltenberg who would have to be persuaded of the UK's good intentions before agreeing to join them in enriching uranium with gas centrifuges.

There were also commercial risks. The British Secretary of State for Defence, Denis Healey, pointed out to Prime Minister Wilson that Europe had to accept that "the United States already has an enormous capacity in their diffusion plants from which they could, if they decided that it was in their own interests, undercut any commercial advantage which the first generation of centrifuges might promise to give a European developer in cheapening the cost of low enriched uranium."[46] Sir Solly Zuckerman was not moved by that argument. For him, commercial considerations were not the only or even the main reason for Europeans' developing centrifuge enrichment. He put it this way to Prime Minister Wilson:

> The Americans have a virtual monopoly of the supply of enriched uranium to the western world. Through this monopoly they (a) arouse apprehension that the enriched uranium that they supply for civil nuclear reactors for generating electricity might be made subject to political strings; (b) endeavour to support the sales efforts of American reactor suppliers; (c) strangle the development of nuclear propulsion for naval purposes by other countries.[47]

Here was the main justification for a European gas-centrifuge enrichment project, even if it was commercially undercut by the United States. Breaking the American monopoly was a "highly attractive" target for Europe, one that other members of the Six would pursue with or without the UK and one that was "essential for [the UK's] European political/economic interests."[48]

Two matters of principle had to be clarified with the US. First, the UK had to be satisfied, and had to satisfy the US, that embarking on a gas-centrifuge project just when the Nuclear Non-Proliferation Treaty was about to be signed would not render the text a dead letter. This concern applied generally (the *N*th-country problem) and to Germany in particular. Second, the UK had to secure the AEC's and the JCAE's permission to share restricted knowledge acquired while British nuclear engineers had collaborated with teams at Oak Ridge between 1960 and 1965.

The Risk of *N*th-Country Proliferation and Its Effect on the Ratification of the Non-Proliferation Treaty

Early in April of 1968, a small British contingent went to Washington to secure assurances that the United States had no political objections to the tripartite arrangements for a commercial gas-centrifuge enrichment plant that were beginning to take shape, and to discuss the general implications of the technological breakthrough for the Non-Proliferation Treaty. The visit anticipated the final round of discussions on the draft text of the treaty, which were to get under way at a meeting of the United Nations General Assembly later that month. The prevailing view in Britain at the time was that centrifuge enrichment probably would not stop any state signing the treaty.[49] On the other hand, British officials were worried that a number of important state parties might later refuse to ratify the treaty, "preferring to keep their nuclear option open, because they were afraid that their potential enemies [would] be able to evade safeguards and manufacture nuclear weapons clandestinely."[50] As the Foreign Office put it, "The prospect of at least twenty countries acquiring a capability to enrich their own uranium within the next ten years introduces a new factor into the non-proliferation syndrome." That risk was all the more significant because "a small centrifuge plant for enriching uranium would be far easier to conceal than the diffusion plants that have been used in the past" and could easily evade safeguards.[51]

The British mission was sent to Washington early in April to prepare the ground for a follow-up visit by a specialized team.[52] It returned home empty-handed. In a lengthy formal statement, the Department of State assured the British that throughout the negotiations on the Non-Proliferation Treaty the US had been asked on numerous occasions what the effect of adhering to the treaty would be "on the ability of signatories to undertake development of the gas centrifuge process, or other processes, for separation of uranium isotopes and to build plants for this purpose if they chose to do so." The US position was "that the Treaty would not in any way inhibit these activities, provided that the output of any plant so constructed was used exclusively for purposes permitted by the Treaty and was subjected to the safeguard provisions of the Treaty."[53] An urgent meeting before the UN gathering was, therefore, unnecessary. Indeed, if it became known that such a meeting had taken place it might create confusion in the minds of those who intended to sign. Nothing should be done now to cast doubt on the adequacy of the Non-Proliferation Treaty and its associated safeguards, once they had been worked out, and to undermine confidence in its security guarantees.[54]

Officials in Whitehall were frustrated by this response. Washington seemed to take it for granted that no government would cheat. With memories of the recent Arab-Israeli War still fresh in everyone's mind, British officials pointed out that "we might find that some of the countries whose participation in the Treaty was most desirable, such as the U.A.R. and Israel, refused to ratify because each suspected that the other might succeed in evading safeguards and manufacturing nuclear weapons clandestinely."[55]

The efficacy of safeguards was the main issue dividing the parties as regards Nth-country proliferation. For the UKAEA, non-intrusive or perimeter safeguards—that is, measurements made outside production plants that checked "what goes in one end and out the other, without any need to look at what is happening in between"—could never properly monitor U^{235} concentrations.[56] However, the alternative, entering the plant itself, though technically far more effective, was politically impossible. Many countries, the FRG among them, feared that on-site inspections might serve as a pretext for commercial espionage. There seemed, then, to be no reliable way to keep countries from evading safeguards if they were determined to do so.

The AEC was far more relaxed about the efficacy of safeguards. The American authorities admitted that perimeter techniques were not particularly reliable when applied to the traditional gas-diffusion process. However, they claimed that that was "largely a result of the inventory already present in such facilities." With centrifuges, by contrast, there was "every reason to believe that similar perimeter techniques" would be more effective. Indeed, the AEC was currently engaged in "additional technical studies directed towards inspection techniques for gas centrifuge plants," and results from those studies would be duly communicated to the International Atomic Energy Agency.[57]

Officials in Whitehall had to navigate between the gloomy prognosis provided by their own researchers in the UKAEA and the upbeat approach to safeguards taken by the AEC.[58] After some hesitation, they came down on the side of the latter. "We have to remember," Christopher

Audland said, "that the proposal for European cooperation on the centrifuge is of very great importance for Ministers in relation to our approach to Europe." Such cooperation was badly needed "to give colour to [Britain's] technological image" on the Continent. If the tripartite project was to be realized, Audland went on, it was "surely essential that we should do nothing that might make the Americans try to slow it down."[59] That is precisely what would happen if the British were to persuade the Americans that the difficulty of safeguarding centrifuges posed a serious security threat. On balance it seemed better to suppress proliferation anxieties so as to allow political and commercial opportunities to blossom, trusting in the American views about the efficacy of the safeguards that would be negotiated once the Non-Proliferation Treaty came into force.

The British government made one more effort to share its anxieties regarding clandestine enrichment with the United States. Prime Minister Wilson raised the issue with Richard Nixon in 1969 soon after Nixon's inauguration as president. They were in a car on the way to London airport when, in the words of an accompanying official, "the Prime Minister referred briefly to the problems which the centrifuge project created in the context of the Non-Proliferation Treaty with the possibility of do-it-yourself plants capable, in the case of very small quantities of enriched uranium, of being set up 'in a garage or a cellar.'" Nixon was unmoved—"It was clear to the Prime Minister that President Nixon was not at all in the picture about this subject."[60] This indifference to the dangers of proliferation was to be one of the defining characteristics of the new administration after more than ten years of concern with what President Kennedy had called the *N*th-country problem.

The Risk of Proliferation by the FRG

The Federal Republic of Germany posed a different problem. Many senior German officials deeply resented a treaty that, if adopted, would condemn their country to second-rank nuclear status and, they feared, authorize foreign inspectors to spy on their civilian nuclear power facilities. Franz Josef Strauss, Minister of Finance in the new post-Erhard coalition government, implied that his country was still being made to pay the price for having lost World War II. He described the Non-Proliferation

Treaty as "a new Versailles, and one of cosmic dimensions too." A furious Chancellor Kurt Kiesinger decried the treaty as indicative of the "atomic complicity of the superpowers."[61] True, in 1954 Chancellor Adenauer had formally renounced the development of nuclear weapons on West German soil. To some American officials, the more the Germans objected now to the terms of the Non-Proliferation Treaty, the more fragile Adenauer's guarantee appeared to be. Glenn Seaborg went so far as to suggest later that the objections in the FRG to the treaty were "all in some degree rationalizations for a reluctance to formally forswear the possibility of ever acquiring nuclear weapons."[62] "If there is one government on earth that cannot abstain from the treaty," the *New York Times* editorialized in July of 1968, "it is the West German government."[63] In fact it took the election of a chancellor who was willing to smooth over relations with Washington—Willy Brandt—for the FRG to announce that it would sign the Non-Proliferation Treaty, and then only "after a vitriolic Bundestag debate that produced a narrow majority in favor" in November of 1969.[64]

To justify a joint venture with the FRG, London relied heavily on the argument that close collaboration would contain proliferation, not accelerate it. West Germany would develop centrifuge technology with or without British help. Any threat it might pose to the existing balance of nuclear power would "be greatly reduced if the collaborative arrangements are such that we are in a position to keep a close check on what is being done in Germany."[65] The transparency provided by having direct access to what the Germans were doing in a tripartite venture under tripartite management and tripartite political control added levels of surveillance beyond safeguards. Anyway, in the initial program there would be no enrichment plant on German soil, and the FRG would not be supplied with centrifuges or with enriched uranium.[66] In fact, Whitehall hoped that "we should be able to make use of the special relationship we shall be establishing to put gentle and discreet pressure on the Germans to sign the Treaty sooner rather than later."[67] Tripartite cooperation, not isolation and with it the strong possibility of an autonomous German enrichment plant, would further the goal of non-proliferation.

France's concerns about granting the FRG access to enriched uranium forced Nixon to engage with the tripartite project shortly after his visit to London in February. His National Security Adviser, Henry Kissinger, sent a formal reply to a complaint that had been made on behalf of de Gaulle by the president of the French National Assembly on March 20, 1969. Kissinger's response confirmed the Department of State's view that the scheme was deserving of support because it would encourage cooperation between the UK and the Continent, and that it was better to see the Germans involved with the British and the Dutch than to see them go it alone. De Gaulle, not satisfied, raised the subject of the FRG's "driving ambition" again at Eisenhower's funeral, ten days later. France was haunted by memories of aggression and occupation going back a hundred years and was irritated by a tripartite scheme that might supply the FRG with enriched uranium for a nuclear weapon. Nixon ignored the complaint. Seaborg, when approached by the French Atomic Energy Commission with the same objections, reiterated Britain's view. By locking the FRG into the venture and by facilitating a close monitoring of its progress, the tripartite agreements would curb proliferation, not promote it.[68]

The United States was then supportive of the tripartite agreement. It dovetailed with overall American policy goals in Western Europe. It did not increase the risk of proliferation, notably to the FRG. On the contrary, Britain's plan involved just the three countries with which the AEC had agreed to classify information about gas-centrifuge technology in the early 1960s. As the official American response to enquiries from Whitehall put it, "we have never constructed our agreement [with these three partners] to preclude classified cooperation and we could not object if some such effort was undertaken."[69] It did not matter much to the US that each partner would contribute its current design of a production model, its experience and technology in plant design, and, in Britain's case, its knowledge of uranium hexafluoride processing and engineering.[70] What mattered was that the knowledge was restricted to the countries that had agreed to keep it classified. Indeed, the United States' biggest concern about the British joint project was not that it would undermine the Non-Proliferation Treaty but that it would proliferate restricted information that British researchers had acquired from their American partners when the two had worked together between 1960 and 1965. Much to the UKAEA's surprise, it was here, at the core of the "special nuclear relationship," that the biggest threat to Britain's ambitions for exploiting the political and economic opportunities of the gas centrifuge emerged.

Collaboration, Competition, and Retardation: Preserving the Model T

The terms of the collaborative research done by British and American engineers at Oak Ridge were defined by Article IX(c) of the 1955 US-UK agreement on civilian uses of atomic energy, which stipulated that "no material, equipment, device or restricted data" and "no equipment or device which would disclose any restricted data" could be passed by either country to a third party "without the written consent of the party to this Agreement … from which it was received."[71] In practice this meant that the British had to reassure the Atomic Energy Commission and the congressional Joint Committee on Atomic Energy that no restricted data of American provenance was embedded in the centrifuge that they intended to pool with their Continental partners. Only then could they proceed with detailed negotiations over the technological contribution of each party to the joint effort.

This priority was brought home forcefully to the British during discussions that got under way in Washington in December of 1968. A high-level delegation had come to the US to discuss the substantial progress made by the UKAEA and its implications for the structure of the tripartite European centrifuge enrichment program. Senior members of the UKAEA and the AEC were accompanied by representatives of several government departments (the Ministry of Technology, the Cabinet Office, the Foreign and Commonwealth Office, the Department of State, and the Arms Control and Disarmament Agency). There were two main topics for discussion on the table: declassification and the sharing of knowledge between the UK and its future partners.[72]

The two teams agreed to disagree on the question of classification. The British argued that it would be necessary to declassify gas-centrifuge technology as industry became increasingly involved, and as more and more countries wanted to exploit the process. They emphasized that eventually they would have to divulge performance characteristics (and not only economic data) to prospective clients in order to take advantage of the commercial opportunities that were bound to arise. They also insisted that substantial declassification could curb runaway proliferation, insofar as a country was more likely to intensify its indigenous efforts on centrifuge development the more the tripartite collaboration backed away from offering future assistance under acceptable conditions.[73]

The Americans were not persuaded by the British argument. They wanted the tripartite collaboration to proceed on the existing classified basis "for as long as possible so that the process of bringing the N.P.T. into force on a broad geographical basis can be taken as far forward as possible before any significant measure of declassification occurs."[74] Even then declassification should be gradual, and if any other country wanted to join the European scheme (Italy was considered a prime candidate) it would have to accept the classificatory regime then in place.

The Americans then wanted to know whether any restricted information derived from the AEC-UKAEA centrifuge collaboration between the end of 1960 and mid 1965 would be shared with the Dutch and the Germans. The British were largely unprepared for this question, and had enormous difficulty persuading the American authorities that they were not in breach of the 1955 agreement.[75] The UKAEA, with strong support from Chief Scientific Adviser Solly Zuckerman, argued that there was no breach of Article IX(c), since their Mark I prototype contained no restricted data of American provenance. The engineers and officials of the AEC insisted that, on the contrary, it did contain such data.

To resolve the dispute, two engineers from the AEC (Paul Vanstrum and Myron Kratzer[76]) and two from the UKAEA (Ned Franklin and Don Peirson) met in London in March of 1969 and in Washington in May.[77] There were bruising, unresolved exchanges over what crucial information the British had acquired from the Americans on the fabrication, dimensions, and coupling of the rotors or spinning bowls that made up the body of the centrifuge. The UKAEA did admit that two features of its centrifuge that had been discussed with the Americans departed significantly from Gernot Zippe's original unclassified design.[78] One was the lower bearing, which had been simplified for economic reasons. The British had adapted an American concept, and used a system of three pillars to support the spinning rotor, rather than the ball-and-magnet device preferred by Zippe. Their innovation had been patented, and was unclassified. The other feature, far more controversial, was the so-called end cap.

The length of the body of a centrifuge contracts when it is spun at high speed (the Poisson effect). To compensate automatically for this deformation of the rotor, the UKAEA engineers fitted a cap over the end of each cylinder. This "lid" was not flat but slightly conical or "dished." The flattening of the cone at high rotational velocities offset the shortening of the cylinder, keeping it aligned between the upper and lower bearings. End caps hadn't been foreseen by Zippe. The UKAEA admitted that the idea of dishing the end caps had been suggested to them by the Americans during their collaborative phase, but only in passing—"in fact during a fifteen minute conversation."[79] Thereafter the British had worked out the details themselves, so they said. They had not had access to secret American reports describing the theory. Instead they had independently drawn on knowledge that was "well known in other fields of high speed rotating machinery" to compensate for the Poisson effect, and had applied it to the particular case of the gas centrifuge. Furthermore, their design had evolved as they had adapted their prototype for mass production.

The American engineers refused to accept this interpretation. Whereas the British spoke of a fifteen-minute conversation, they spoke of "a great deal of discussion." Whereas the British dismissed the significance of the exchange by claiming that no "unique" design feature had been transmitted to them, the Americans insisted that "it was difficult to conceive that some of this discussion was not reflected in the ultimate UKAEA design." Indeed, Vanstrum and Kratzer remarked that, in the midst of the collaboration, the US had taken out a classified patent in Britain on the end-cap solution. How then could the British suggest that this could be readily devised by any competent engineer? The patent application was "evidence that the United States regarded their 'end-cap' solution as unique and patentable as early as 1962 and that it was important to the centrifuge technology." Of course the use of a conical cap to compensate for the Poisson effect at high rotational speeds was generally known. "It was the linking of this simple design solution to this particular problem which was important," said Myron Kratzer.

The exchange with the AEC on March 4 persuaded the British that their attempt to dismiss as insignificant any knowledge acquired during the earlier collaboration had failed. The next day, Zuckerman proposed to go further and to provide the Americans with a "statement" that "would specify, in general terms, the materials, the general dimensions of the centrifuge, and the general nature of its bearing and end-cap." This document would also explain why the UK did not believe that any information in their centrifuge could reasonably be regarded as "restricted data" in terms of Article IX(c). AEC Commissioner Gerald Tape was not satisfied. He wanted Ned Franklin to describe the Mark I prototype "in greater detail."

He wanted specifics. In fact he thought that it might be a good idea if the UKAEA arranged for US personnel to "see the United Kingdom gas centrifuge project at first hand."[80]

The British were deeply distressed by the demand for "visual access." Their Mark I centrifuge was the "heart of [their] current contribution to the proposed tripartite collaboration." The life-tested design of the end caps that they now used on the Mark I centrifuge gave them a technological advantage over both the US and their European partners. (The immense forces generated on the walls of rotating centrifuge bowls at high velocities limited their lives and a great deal of research was done to secure their longevity). If crucial knowledge about this design were withheld to please the Americans, the entire Anglo-Dutch-German project would be thrown into disarray. "A 'dished end cap,'" Zuckerman wrote, "should not be allowed to become a barrier to a major European political policy which the American government has not only endorsed but also encouraged."[81]

Many in the Cabinet were convinced that the main reason for the American demand for visual access was not to protect national security but to see what technological improvements the British had made since the collaborative effort had been unilaterally suspended by the US in June of 1965. It mattered little that the Americans assured London that they had no intention of replacing their gas-diffusion plants with centrifuges. Zuckerman put it this way to Prime Minister Wilson: "The Americans are out to dominate the world market for nuclear fuel. Were we to allow them access, they might well pick up ideas from our production model which could make a real difference to their commercial exploitation of the centrifuge in third countries, if not in the USA."[82]

Little did Zuckerman know that the commercial issue had already been settled in principle. Soon after the meeting in March, a team from the AEC and the Department of State saw blueprints of the UKAEA's gas centrifuge. The engineers detected "key bits of information which exposed the primitive nature" of the machine. Indeed, after the meeting, as a Department of State official put it, "our experts referred to the British machine as a 'model T'" and "no longer feared any real competition in the fuel enrichment business from the tripartite collaboration for the next few years."[83] This did not solve the issue of whether or not the "Model T" incorporated restricted knowledge. But it did defuse the issue, at least for the Americans. In deference to the policy of retardation, their technical judgments were not passed on to their embattled partner.

On May 3, 1969, Zuckerman went to Washington to present the British case against visual access to the AEC.[84] Chairman Glenn Seaborg and Commissioners James Ramey and Francesco Costagliola were present. Gerald Tape was also in the room, although he had officially left the AEC just a few days before. The United States' position had hardened considerably. (We do not know whether or not the Commissioners were aware that the British design was "primitive.") Ramey quickly made it clear that the AEC had to have visual access to the British Mark I production model in order to be able to reassure the Joint Committee on Atomic Energy that no American information would be passed to the Dutch or the Germans. What did "visual access" mean? Tape and others explained that it "would not involve taking the model apart screw by screw and item by item," but it would mean "more than just a visual view." "Experts who knew something about the technology [would have to] examine it." This investigation would not extend beyond looking at an individual Mark I centrifuge. The British would not be asked to divulge details of their life-tests, nor would they have to explain how they engineered the device for mass production.

That was not all, though. Ramey insisted that there be a "firm procedure for informing the Americans in advance of what was to be transferred to the Dutch and Germans by way of advanced R and D for future models"[85] so that the AEC could check that no restricted data of US provenance were being shared in the subsequent development. Tape put a time limit on the constraint; he surmised that, as research progressed, by 1980 it would be "hard to find a connection with the 1960–65 exchanges." This was roughly the year in which the AEC itself was planning to commercialize centrifuge enrichment.

A Cabinet Committee chaired by Zuckerman discussed the implications of the US demand for visual access with Chairman John Hill of the UKAEA, and with one of the UKEA's senior engineers, D. G. Avery, on May 20.[86] They realized that visual access to the Mark I centrifuge did not mean a quick look. "The Americans would expect to be able to take our machine apart and to examine the pieces." Because both teams had begun with Zippe's basic design, and because the idea of the end cap had been suggested by the researchers at Oak Ridge and had been discussed with them "many times," a detailed examination of the British device would "inevitably" reveal a "large number of points of similarity." The AEC's "inspectors" would be accompanied by lawyers who would emphasize these points of similarity (rather than the distinct contributions made independently by the UKAEA engineers), leading to "further argument and recrimination." Hill concluded that, on balance, it was better to produce a reasoned argument against visual access than to open the prototype to inspection and give the JCAE "more information with which to cause trouble."

Hill was asked to dig up the records and to outline the argument. His findings were conveyed to Zuckerman the next day. Zuckerman had just penned a memo claiming that the British end cap was based on an independent application of a general engineering principle. "No documents or 'blue-prints' were exchanged on this specific issue," he had written, so that "it would be immensely difficult to distinguish between specific US classified information and our application of basic principles of physics and engineering."[87] Hill's new findings were a bombshell. The concept of the end cap had first been mentioned in 1962. In 1963 the US had disclosed the dimensions and the performance of an end cap 4 inches in diameter. In 1965 "two detailed papers were presented by the US dealing with the stress analysis and lift of a dished end cap and reporting on their prototype."[88] The main difference between what the British had learned in these exchanges and what they did subsequently came down to developing a rotor with a diameter of 5 inches rather than 4. Secretary of State for Defence Denis Healey immediately demanded an official enquiry.

The enquiry, led by Lord William Penney, reported ten days later. Its report confirmed that at an "especially significant meeting" in May of 1963, American engineers "showed the British team a drawing of their deforming end cap for a 4 in. diameter cylinder, together with experimental results on the deformation of this type of cap at various rotational speeds for various angles of the dishing cone and for various thicknesses of metal." The report went on to say that "they also told the British team of their theoretical work on the stresses and deformation of the end cap, and showed a drawing which included a dished baffle plate" (that is, dished diaphragms within the rotating cylinder that served as gas baffles). The principle, the proportions, and the material of the dished end cap on the Mark I centrifuge were the same as those described to the British team by the Americans. Its geometry incorporated only "minor modifications" of the US model.[89] The encounter had not been limited to a brief conversation. The UKAEA had learned about the end cap in the course of a "fairly lengthy exposition, followed some two years later by the transmission of US papers."[90] The exchange went well beyond basic research. It extended "from laboratory experimentation up to and including pilot plant design and operation."[91] At the meeting with representative of the AEC in London in early March, Myron Kratzer had insisted, against UKAEA protests, that in terms of Article IX(c) of the 1955 agreement "the United Kingdom could not include the conical end-cap in any exchanges with the Netherlands and Germany" without explicit permission from the United States.[92] Penney's three-man panel concluded unequivocally that the design of the British end cap incorporated restricted data of US provenance. Kratzer was vindicated.

On June 3, Sir Solly Zuckerman was back in Washington to meet with representatives of the AEC and of the Department of State. His hands were still tied by a UKAEA that was deeply reluctant to let American researchers into British labs. Zuckerman proposed to show the Americans an end cap and other features of the Mark I centrifuge that they thought might have been derived from their research, but not to grant them visual access to an entire centrifuge. The UKAEA's representatives also offered to give the AEC an "outline" of what they intended to share with their Continental partners, and to consult with US engineers if and only if the British deemed it necessary.[93]

Both proposals were summarily rejected. Nothing short of visual access to the entire Mark I centrifuge would satisfy the JCAE. Visual access would also be necessary in the future, if the AEC requested it, to check whether any American restricted data had been used in the designs of later production centrifuges. The UKAEA capitulated. In an aide-mémoire dated July 3, 1969, Her Majesty's Government agreed "to grant to the US authorities visual access to their first production model." As regards future R&D, HMG agreed to provide the US authorities with "an outline of the original UK contribution to the initial collaborative Research and Development programme" with the Netherlands and Germany. HMG also agreed to consult in advance with the US before transferring information to Britain's partners in the tripartite enrichment venture. This consultation on future developments would include "access, if requested by the US authorities, to the specific data or features" that the Americans suspected might violate the terms of Article IX(c) of the 1955 agreement.[94]

A four-man American team visited UKAEA facilities at Capenhurst and Risley for two days in mid July to inspect the centrifuge program. Led by Commissioner Theos Thompson, the team included Paul Vanstrum (a member of the centrifuge team at Oak Ridge), one other centrifuge expert, and Nelson Sievering from the Department of State.[95] The UKAEA scientists and engineers cooperated fully. Sir Solly informed Prime Minister Wilson of the results of the visit. Thompson had told Zuckerman that his team had been impressed by the amount of "independent development work done in the United Kingdom." The discussions and presentations had dispelled "various technical doubts raised in the US."[96] Thompson was back in England early in March of 1970. He spent two days "to satisfy himself personally" about those aspects of the British centrifuge program which the UKAEA would disclose to their two Continental partners. Just before leaving, Thompson told Zuckerman that "as far as he [was] concerned, all [was] well." though of course he warned him that the JCAE might deem otherwise.[97]

The dispute was officially closed on October 1, 1970.[98] The AEC officially agreed to let the British government proceed with the tripartite venture even though, in its view, the design of the end cap on the British prototype contained information that was restricted in terms of Article IX(c). The AEC representatives were emphatic, though, that this should not set a precedent for further technological exchanges in the joint venture. They also took note of British assurances that this case had "no relevance to the exchanges of information with the United States in the fields of nuclear weapons and of naval nuclear propulsion under the 1958 Agreement for Cooperation [on the Use of Atomic Energy for Mutual Defense Purposes]." With the path now cleared, the governments of the United Kingdom, the Netherlands, and the Federal Republic of Germany quickly put the final touches on a treaty establishing a new European organization for gas-centrifuge enrichment, which eventually came to be called Urenco.[99]

How did the Atomic Energy Commission hope to satisfy the Joint Committee on Atomic Energy when, in fact, the British centrifuge did incorporate a bottom end cap that resembled "the cap used on AEC's 6[-inch] machine, which the UK scientists saw during the 1960–65 exchange"? Theos Thompson gave the JCAE a number of reasons for dropping the case. Clearly the UK was determined to "preserve the basis of mutual trust" that informed its several cooperative agreements with the US. In addition, the UK had confirmed unambiguously that this case would not be appealed to as a precedent for the sharing of sensitive military knowledge covered by the 1958 agreement. There was also the danger that, if the US now impeded the tripartite venture, the FRG might hesitate to ratify the Non-Proliferation Treaty. The FRG had been "skittish" about how that treaty might affect its right to enrich uranium for a civilian nuclear power program, and an effort by the US to impede the tripartite venture might confirm their fears.

Most important of all, the "actual inspection of the machine" revealed that it embodied a "relatively unadvanced state of the art" in comparison with the current American centrifuge.[100] In fact, "although the British [did] not know it, the end cap in question [was] no longer used in the advanced AEC machine." If the JCAE insisted that the UK abolish the end cap now, it was more than likely that they would come up with an improved design that wouldn't require an end cap, as in current US models. It would be far better to let the UKAEA persist with its poor prototype than to "force them to a more advanced model earlier than their present schedules would seem to call for. Because of the substantial investment which the UK has in the Mark I machine they would probably stick with it for at least five years. From the standpoint of proliferation and potential competition with the US in the foreign uranium market, there is some advantage to leaving things as they are."[101]

The AEC was, of course, under no obligation to inform the British that the design of their prized prototype was "primitive" in the view of its centrifuge engineers. The AEC's task was to report to the JCAE as to whether or not its British counterpart had respected the terms of the formal agreements between the two governments. The British had not, clearly. But it was easier to be forgiving if the device that they wanted to share with the Germans and the Dutch was a generation or more behind the state of the art. Allowing the British to persist with their "Model T" also ensured that American pre-eminence in centrifuge technology would not be challenged. Collaboration easily morphs into competition. Rivalry surfaced now clothed in the language of non-proliferation.

Concluding Remarks

In his fine overview of US-UK nuclear collaboration in the wake of the 1958 amendment to the McMahon Act, John Baylis is at pains to emphasize that the "special (emotional) relationship" between Britain and the United States had little if anything to do with the willingness of both sides to share restricted knowledge on atomic weapons. Collaboration was driven, rather, by reciprocity and mutual respect—by the recognition that each side had something of value to bring to the table. We see little of that here. Indeed, the abiding impression that one has of the UK-US relationship described in this chapter is the suspicion that major British stakeholders had of American intentions, and the fear of retaliation should the British step out of line. Technological collaboration did not take place on a level playing field, as Baylis' "reciprocity" implies, but in an asymmetric frame with the US at the dominant pole. Five years of joint effort to improve the centrifuge, and the UKAEA's decision to exploit the technology commercially, turned Britain from a desirable partner into a potential rival. UKAEA research scientists and engineers wondered whether their American colleagues deliberately misled them in 1967 by playing down the commercial potential of the centrifuge. Whitehall did not press its concerns over *N*th-country proliferation at the dawn of the Non-Proliferation Treaty for fear that the AEC might change its mind and put obstacles in the way of the tripartite venture. Secretary of State for Defence Denis Healey was convinced, and not without reason, that major military cooperative agreements entered into with the United States after 1958 (e.g., for the provision of nuclear submarine fuel, or for the exchange of weapons information) might be jeopardized if the UKAEA did not collaborate fully with the AEC and the JCAE.[102] And then there was the question of the end cap.

Both Sir John Hill and Sir Solly Zuckerman believed that the American authorities were making a mountain out of a molehill as regards the end cap—Zuckerman described it twenty years later in his autobiography as a "minor feature in our design that had been discussed at a low-level technical meeting between experts of the two sides."[103] Both were rightly convinced that the question of the end cap had been used as a pretext to gain direct access to the British prototype. Apparently an end cap wasn't even used any longer on American centrifuges. In fact, in Nelson Sievering's view "the poor potential of the UK production model" was due in part to the end cap. Visual access to the British centrifuge amounted to industrial espionage under the guise of security concerns. It was an act driven by the spirit of competition with a rival rather than the spirit of collaboration with an ally. Determined to retain US technological superiority, and to control the flow of enriched uranium on the world market, the engineers at Oak Ridge alienated a major ally to satisfy the exigencies of the JCAE. Their power to extract concessions from the UKAEA, to the point of having visual access to the Mark I centrifuge, derived from the web of technological and political interconnections that tied the US and the UK together in the nuclear field, beginning with classified collaboration on the centrifuge between 1960 and 1965. The "special relationship" provided the UK with privileged access to the classified research system, but it also ensnared the UK in a web of constraining obligations. The ensuing bitterness in London was summed up by Minister for Technology Tony Benn after he learned of the conditions the AEC had imposed in exchange for overlooking the UK's breach of the 1955 agreement:

> What came out of the meeting, which I had suspected but had never been properly told, was that we have an arrangement with the Americans under which we are absolutely tied hand and foot to them, and we can't pass any of our nuclear technology over to anybody else without their permission. The harsh reality is that de Gaulle is right.[104]

Technological collaboration, fanning out over a wide spectrum of US-UK nuclear relations, provided Washington with an instrument of soft power that it could wield to achieve its political and economic objectives almost at will. It is these linkages that "hardened" the backbone of soft power—linkages that the US administration had trouble establishing in the other cases discussed in this book.

Conclusion

This book has foregrounded the role of American-led technological collaboration (or denial) as an instrument of soft power, promoting European integration and curbing the proliferation of nuclear weapons and their delivery systems in the early and middle years of the Cold War. By securing US control over the military and commercial exploitation of enriched uranium, and by seeking to steer limited European resources down civilian research trajectories, these initiatives dovetailed with the broader foreign-policy objectives of successive administrations, helping to consolidate a "third great force" (Eisenhower) at the heart of Europe without threatening US leadership or destabilizing the balance of power in the region.

The role of technological collaboration in promoting the spread of nuclear weapons has always excited controversy.[1] It has been given an added urgency by the emergence of "rogue" nuclear states, such as North Korea and Iran, that have capitalized extensively on foreign technological assistance to build their nuclear programs (though of course only one of them has so far acquired nuclear weapons). This book contributes to the debate on non-proliferation strategies by describing the contested, evolving policy-making process in Washington regarding technological sharing or denial whereby the administration sought to pre-empt proliferation by fostering institutionalized arrangements and technological options calculated to retard, if not stop, the political decision to develop a nuclear weapon.

Seen from the point of view of the circulation of technology and know-how, the central dilemma of the Cold War was that the core technologies that contributed to the arms race between the superpowers were "dual-use"—that is, in important respects they simultaneously had both military and commercial applications. This produced a tension between state regulation and market penetration. It also generated a sense of inevitability to the dissemination of certain technological components that were necessary to build nuclear weapons and the missiles that delivered them to their targets. Neither the production of nuclear power nor the development of satellite launchers could be halted in an economic system committed to protecting the free reign of market forces in the civilian domain. The states that developed them indigenously acquired cadres of trained experts and the industrial capability to produce fissile material and rocket propulsion systems that could, if the need arose, be diverted from a peaceful program into a military one. Technological collaboration was invoked to nudge them away from that path.

As I emphasized in the introduction, the concept of technology used in this book refers to an artifact that has meaning only in a broader technological system. Curbing proliferation involves intervening at many sites (some more crucial than others) in an interconnected web of loosely coupled, interdependent components that together constitute a "technological system-in-the-making." For my actors, thinking about proliferation involved controlling the production and the circulation of technologies as diverse as fissile material, rocket fuel, high-performance computers, warheads, and guidance systems, all embedded in a network subtended by major investments in financial, industrial, and manpower resources. Technological collaboration or denial are among the many instruments the United States had at its disposal to curb proliferation of very different kinds of sensitive materials. Others included classification, export controls, visa policies, international safeguards, international treaties, and less formal normative regimes that impose non-bellicose technological choices on signatories. (Today there are more than 2,000 items on the US Commerce Control List that regulates the circulation of dual-use technology.) If technological collaboration stands out, it is because it offers something in return to a state that agrees to align its technological trajectory with the wishes of the dominant partner.

The success of this strategy depended on US pre-eminence in many fields of science and technology. The knowledge gap between American researchers and their British and European colleagues was a political lever that Washington could use as a bargaining chip that served to achieve broader strategic objectives. The US did not primarily require a mutual exchange of knowledge across the Atlantic, as was the case in US-UK exchanges of information on thermonuclear weapons beginning in 1958.[2] Calibrated technological collaboration was a multivalent strategy that served many purposes, catalyzing (so it was hoped) and then consolidating the process of European integration, closing the technological gap, creating markets for US corporations, and curbing proliferation.

By the early 1960s, technological collaboration was not intended to curb proliferation completely. After the French and Chinese nuclear tests in 1960 and 1964, respectively, the US realized that it would be difficult to stop any state that wanted to build nuclear weapons from doing so. This risk would be even greater once production-scale gas-centrifuge enrichment had been mastered. But if the US could not stop them, it could at least slow them down. Atomic Energy Commission Chairman Glenn Seaborg advocated a "prevention or maximum delay policy" to impede *N*th-country proliferation.[3] Classification, technological denial, or technological sharing to divert resources away from military into civilian programs all served the more limited goal of a "delay policy." They could postpone the decision to embark on a weapons program, and in so doing they could allow more time for diplomatic and other measures (such as a change in government or leadership or the emergence of a strong local anti-nuclear movement) to take effect.[4]

Technological collaboration was just one of a number of instruments at the United States' disposal to advance its interests in Europe. Its effects cannot be evaluated without taking into consideration the broader strategic context in which it was implemented. Indeed, a great deal of social work was required to implement policies that exploited the United States' technological pre-eminence against countervailing forces that believed it was crucial to maintain leadership at all costs. This study has emphasized that curbing proliferation in practice pitted different arms of the US administration against one another, mostly because they disagreed on what counted as "sensitive" assistance to another state—that is, assistance that would be to the detriment of the United States' military and economic security. Consensus had to be brokered between multiple agencies with different agendas, including technical agencies such as the AEC (which was beholden to the congressional Joint Committee on Atomic Energy) and NASA, the Department of State, the Department of Defense, the Arms Control and Disarmament Agency, the White House Office of

Telecommunications Policy, and Congress. These internal disputes were sometimes bruising and often led to results that were less than desirable from the viewpoint of the Department of State—and of its European partners. Congress' insistence on applying a Buy America rule to the Export-Import Bank's loan to the joint venture with Euratom, against the Department of State's plans, struck a blow against European enthusiasm for the project, which was further dulled by the declining commercial prospects for nuclear power. NASA and the Department of State were particularly frustrated by the Office of Telecommunications Policy, which, to protect the single global satellite communications system and Comsat, hollowed out the technological content of the offer to the European Launcher Development Organization. In this case, in fact, the conditions imposed by the US were so restrictive that Europeans eventually decided that they would rather continue without technical help than find their launching programs disrupted by Washington.

Did technological collaboration achieve the desired results? Was it a useful instrument of soft power? The case studies discussed in this book have exposed its limits rather than celebrated its success. The joint program with Euratom was an admission that officials had overestimated the United States' lead in the development of civilian nuclear power, as Robert Schaetzel of the Department of State openly admitted. Though the success of positive disarmament in the FRG cannot be determined, it is not likely that a "diversion" of resources into space science had any impact on the domestic debate on the FRG's right to acquire nuclear weapons. Certainly it did nothing to quell widespread opposition in the Bundestag to the government's signing the Non-Proliferation Treaty in 1969. The attempt to inject new life into ELDO, and to keep the British involved, did nothing to save it from ignominious failure. Not only did London have no interest in technological support; ELDO's fate was almost sealed from the start anyway— not for a lack of technology, but for a lack of a centralized system of project management that could ensure the integration of the three rocket stages built in separate countries. In each of these cases, US officials placed far too much faith in technological leadership and underestimated the power of national interest in European governments. In major areas of nuclear and space technologies, European governments sought support from the United States to strengthen their autonomy and found themselves drawn into a dependence that they did not want. This tension was always inherent in the American project for rebuilding an integrated postwar Europe. McGeorge Bundy once asked rhetorically why the US sought a strong, united Europe, "since great states do not usually rejoice in the emergence of other great powers."[5] His answer—their shared interest in resisting communism—was, for many in the administration, not reason enough to go along with the policy. The limitations of technological collaboration as a tool of soft power were due as much to the exaggerated faith it placed in technological solutions to political problems as to the nature of the technologies involved, the high stakes, and the inevitable internal disagreement that they engendered.

The balance sheet is not entirely negative. The early enthusiasm for Euratom of President Eisenhower, Secretary of State Dulles, and the Department of State (with wavering support from the Atomic Energy Commission) undoubtedly catalyzed the more ambitious program of the Six (France, the Federal Republic of Germany, Italy, Belgium, the Netherlands, and Luxembourg) to forge a European economic community. A nuclear power technological platform on which to build a supranational Europe that would envelop German nationalism and militarism and curb proliferation combined "atoms for peace" as a resource for developing atomic energy and as a "disarmament" strategy with the construction of a United States of Europe, all major foreign-policy goals. Its effects on dissuading the FRG from acquiring nuclear weapons may have been more muted than Gunnar Skogmar claims—after all, the governments of France, West Germany, and Italy were secretly scheming to collectively embark on a program to develop nuclear weapons even while the Euratom treaty was being signed and ratified. All the same, it is reasonable to surmise that support for Euratom contributed to creating a climate of opinion that curbed proliferation by embedding nuclear power in the deeper politico-economic process of postwar European integration, and was consolidated by formal measures such as the Nuclear Non-Proliferation Treaty. The enthusiasm of the Department of State and of NASA for ELDO, and their limited offer of technological collaboration, did not save ELDO from its own internal contradictions; it did send a strong signal to the member states, notably Britain, that European integration around civilian technological platforms still mattered in Washington.

Similarly, the support for the tripartite, British-led centrifuge enrichment program was aligned with the Department of State's wish that London draw closer to the Continent and locked the FRG's advanced work on gas centrifuges into a multinational organization. It did so at the expense of a purely national program that would have been far more difficult to monitor.

As I suggested at the end of chapter 5, technological collaboration is more likely to be successfully mobilized as an instrument of soft power if it is one of a number of overlapping ancillary power relationships. Eisenhower's support and his enthusiasm for "atoms for peace" reduced tensions between the AEC and the Department of State over Euratom. It also facilitated the passage through Congress of the enabling legislation for the joint program, and crushed Sterling Cole's and the nascent IAEA's profound opposition to self-inspection. Technological collaboration's success as a political instrument with the UKAEA (the Dutch and the West Germans were far less beholden to the US on centrifuge technology) was possible because it was the stuff of a formal legal agreement between the partners. Breaking that agreement had serious consequences for the UK, whose nuclear program was dependent in multiple ways on American materials and know-how. It was these linkages that enhanced the power of technological collaboration. No such linkages were there in the case of space collaboration—in fact, the standing policy was against sharing missile technology with France or any other power. The White House Office of Telecommunications Policy was unmoved by NASA's and the Department of State's calls for flexibility. If technological leadership is to be used effectively as an instrument of soft power in sensitive domains such as those discussed here, it is best "hardened" by being embedded in a network of mutually reinforcing relationships that enhance its reach. The Gilpatric Report of 1965 recognized that no single measure could be counted on to halt the spread of nuclear weapons. Its recommendations combined general strategies such as international agreements with country-specific cocktails to curb proliferation. Technological collaboration alone might achieve little more than indicating the United States' wish to maintain a foothold abroad in an important strategic sector; associated with other measures, it surely had a role to play in stabilizing the international system.

On Technological Denial

Granted the limitations of technological sharing as an instrument to shape the technological choices of another state "along the lines the dominant state prefers" (Nye), technological denial presents itself as a far more potent alternative. Denial targeted at particularly sensitive elements of a weapons program (e.g., a uranium-enrichment facility) can impose an enormous burden on a government that is obliged to build its own plant to secure its autonomy. At a JCAE hearing in 1960 it was stated that France's gas-diffusion plant as Pierrelatte would cost $750 million to build, an astronomical sum at the time; in 1968, the British Minister of Technology, Tony Benn, derided the plant as woefully inefficient.[6] A "blanket" refusal to share technology can also seriously retard a program for the development of nuclear weapons or of missiles. This happened in France pursuant to the 1964 White House National Security Action Memorandum 294. McGeorge Bundy, infuriated by Charles de Gaulle's refusal to fold his defense policy into that of NATO, signed off on NSAM 294, which stipulated that the US government would not "contribute to or assist in the development of a French nuclear warhead capability or a French national strategic nuclear delivery capability." It listed a number of modes of support that were excluded, including "exchanges of information and technology between governments," and "joint research and development activities."[7] The effects of this "blanket" denial of technology and know-how were revealed six years later. John Foster (a physicist at the Lawrence Livermore Laboratory and the Director of Defense Research and Engineering in the Department of Defense) told National Security Adviser Henry Kissinger, perhaps only half-seriously, that France's program was the worst in the world—even worse than China's.[8] The French themselves estimated that they were fifteen years behind the Americans.[9] Technological denial was obviously effective.

Prudence is called for. First, the very success of technological denial was a cause for alarm, not for celebration. Kissinger was profoundly disturbed by the backwardness of the French military program. Its collapse, he said, would lead to neutralism. Now that it existed, it was important that "it be effective and above all that it does not become irrelevant." Better to help the French develop a credible deterrent system to meet a growing Soviet offensive capability than to leave it exposed and vulnerable.[10]

The United States' denial of technological assistance to France was also effective because it was not undermined by a rival nuclear power. In the 1950s and the 1960s, apart from the Soviet Union, Britain was the only country with a mature bomb program to which France could turn for technological help. In 1957 the French sought assistance with the design and the testing of weapons, with their gas-diffusion plant, and with the provision of enriched uranium. Meeting in Bermuda in March of that year, British Foreign Secretary Selwyn Lloyd assured US Secretary of State John Foster Dulles that "the UK would take the same negative position as the US would take in this matter."[11] That agreement cut off the only other source of advanced technology that France could draw upon. In today's world, with multiple nuclear powers, the risks that a policy of bilateral blanket denial will be subverted by a rival power cannot be ignored.

The circulation of knowledge through commercial channels also frustrates any attempt to hermetically seal off a country from external technological support (leaving aside espionage and the theft of nuclear materials). A report prepared by the Department of Defense for Kissinger put this well: "France along with other countries has had virtually free access, through commercial channels, to much of the broad continuously advancing technological base existing in this country. Much of this technology has peaceful as well as military applications and flows in relatively unfettered—and quite legitimate—channels."[12] A typical example was the high-performance computer sold to France on condition that it be used only for non-military purposes. In practice there was no way of keeping it from being used for the nuclear-weapons program; in fact, everyone knew that it was.

Blanket technological denial is, then, an elusive goal. It also has serious political disadvantages, which were evident to Dulles when he met with Lloyd in Bermuda in 1957. "The US was not disposed to assist the French in a weapons program," Dulles said. "If they wanted to produce nuclear weapons they must do it on their own," he added. The Secretary of State backpedaled at once. "He was not sure that open opposition to a French program was the right way," Dulles mused. It would "arouse nationalistic emotions which might speed up the French program." Better, then, to retain some "minimal association" with the French in this area, "with appropriate 'dragging of feet'" to slow development.[13]

Dulles' instincts were right. In the early 1980s, a political scientist at Princeton University, Richard Ullman, conducted more than 100 interviews with senior French and American officials about their still-secret dealings related to weapons technology in the 1970s. One knowledgeable French observer told Ullman that the policy of denial had stiffened nationalism, provoked anti-Americanism, and fueled a determination to proliferate in defiance of US policy. As he put it, the leading figures in the French weapons program were strong nationalists who had long tended to feel that the United States "was really the enemy" because, by seeking to impede the growth of the French force, it had sought to prevent France from taking its rightful place in the world. In reaction, the Commissariat à l'Énergie Atomique had developed an unspoken bias in favor of proliferation. "Probably because we had not quite overcome the bad memory of having been pushed to the wall by the United States," this individual said, "we were reluctant to join the club of those who push the others to the wall." All this changed when collaboration resumed during the Nixon administration. "From the moment we got information from the US," the interviewee said, "we were more open to talk in a sensible way about how we would handle the overall nuclear situation in the world."[14]

In sum, significant restrictions on technological collaboration to curb proliferation certainly affect a state's weapons program. But they can never be totally comprehensive, they risk being undermined by rival powers and commercial transactions, and they can have very negative political effects. Denial carries dangers of its own. It is not an "obvious" alternative to technological collaboration as a tool of non-proliferation.

Technological Collaboration and the Circulation of Knowledge Between Scientists and Engineers

A great deal of research in the social sciences black-boxes technological collaboration itself and the policy conclusions drawn from it. This is partly because of the sources used (official government papers in my case), which usually do little more than authorize certain transactions and not others. How those transactions are implemented in practice—how technology and knowledge actually flow across borders and are appropriated—is generally not visible in these documents. Firms that ship equipment abroad and train their clients in how to use it (subject to export controls) are actors in this "invisible" process. So too are scientists and engineers, members of the international research community who can monitor the uses to which the equipment is put and can sense whether it is being diverted from peaceful to military uses.

I have emphasized that the technological systems that concern us here are evolving organisms built from multiple, loosely coupled components that are constantly being improved, and that they should be understood as works in progress transformed by new knowledge injected into them. That is what makes "proliferation" a nebulous concept. It alludes to a situation that is difficult to control and that calls for constant vigilance. To conclude the book, I will briefly shift the focus of the analysis onto the practice of technological collaboration itself and discuss some of the general mechanisms by which a technological system is transformed by the transnational circulation of equipment and knowledge.[15] To illustrate the argument, I will use recently declassified material describing the fevolution of US policy to help the French improve the weapons program they had. This is not a story about curbing proliferation, then; by 1970 the US could not stop the French nuclear program in its tracks. But it does throw light on the practice of technological collaboration and on its importance as an instrument of control and as a tool to stabilize world order.

It must be emphasized that technological sharing and technological denial are not alternative strategies. They coexist. Indeed, both may be applied to a high-technology device such as a computer—some models will be freely available, others will be controlled. The same was true of rocket fuels. George Ball was careful to distinguish non-storable liquids from storable liquids and powders. The latter posed a proliferation risk, the former far less so. Of course only highly qualified technical people are able to make these judgment calls. They too are an essential part of the conversation on non-proliferation.

A better understanding of the broader context is needed before I address the substance of the change of heart by Henry Kissinger and Richard Nixon as regards France and the use of technological collaboration as an instrument of soft power. Shortly after Nixon's inauguration as president in January of 1969, he and Kissinger decided to revise Lyndon Johnson's policy of isolation and to assist the French with their weapons program so that it could be better integrated into the defense of Europe. According to Kissinger's recent biographer Jeremi Suri, they made it their goal to "transform the structure of the international system to encourage a diffusion of power on terms favorable to the United States." Kissinger claimed that a world with more "centers of decision"—a more multipolar world—would provide flexibility for innovative diplomacy and consensus building rather than the enforced dominance of reluctant allies.[16] Kissinger struck up a fruitful collaborative relationship with de Gaulle's successor, Georges Pompidou, and at least until 1973 they worked together to improve the capability of the French nuclear system.

Kissinger was quite explicit about his reasons for helping the French and about how he intended to use the United States' technological collaboration as an instrument of soft power. Indeed, his is almost an ideal type of the genre. Kissinger's principal objective in responding positively to France's requests for help was not to exchange technology; the French had nothing the US wanted. Instead his aim was to shape "the basic orientation of French policy," to secure "a substantial French movement back into the integrated Alliance structure."[17] That reorientation would be gradual, and would be managed carefully so as not to harm the United States' national or economic security. In the summer of 1973, Robert Galley, France's Minister of the Armed Forces, visited the "Western White House" to discuss what he needed. Kissinger asked Secretary of Defense James Schlesinger and his Deputy Assistant, Brent Scowcroft, to "lead them on without giving up anything—we want to get a handle on them without [their] knowing it."[18] Schlesinger and Scowcroft had to devise strategies to give the French something that "looks like a step forward but doesn't give them anything yet."[19] Kissinger put it bluntly: "What we want is something which makes Galley drool but doesn't give him anything but something to study for a while. I will brutalize Galley."[20] Thus brutalized, Galley would help Kissinger "break up the Europeans." "We are going to try to bust the Europeans," Kissinger said. "The French can be useful in this."[21] A bilateral agreement with France, he imagined, would eventually create a backlash from Britain and the other Europeans who would turn on them saying "You bastards, you talk about unity and then you go this bilateral route" with the United States. [22] This was a far cry from the promotion of an integrated United States of Europe that had been the cornerstone of US policy for more than 15 years.[23] But Kissinger was determined to use knowledge sharing as a bargaining chip to engage France's desire for independence, both to weaken the power of the growing European bloc (which now included Britain—it joined the European Community in January of 1973) and to integrate the French nuclear deterrent more closely into that of the Atlantic community.

Kissinger's behavior confirms the point made above that in sensitive fields technological collaboration works best as an instrument of soft power if it is consolidated by other means. It was precisely because "classical" supportive linkages were not available (e.g., it was generally believed that Congress would strongly object to sharing knowledge about weapons with the French) that Kissinger had to resort to deceit and bullying to secure his goals.[24] When Robert Galley arrived in San Clemente, for example, the first thing he and Kissinger did was devise a scheme to cover up his trip. I quote the transcript here.

Kissinger: If anything should get out to the press, we'll just say that you stopped here on your way back from Tahiti.
Galley: We took every precaution to ensure secrecy. I am sure no one knows on our side.
Kissinger: The trouble on our side is that the Air Force people see your plane.
General Walters: We can just say the plane landed here because the other field was closed. It was to land at March [Air Force base, near Riverside] but it had to go to El Toro [Marine Corps Air Station, near Irvine] instead.
Kissinger: We'll say you are an old acquaintance, your plane was diverted to El Toro, you called to pay your respects, and I asked you down. There is very little possibility that this will happen, but I like to be prepared.
Galley: The press is not interested in things over weekends.[25]

Subterfuge was coupled with outright threats. Kissinger developed a visceral contempt for one senior French Minister, Michel Jobert, who made particularly hostile comments about US behavior. For a brief moment, circulation of knowledge was jeopardized. Kissinger imposed order. At a meeting of the North Atlantic Council in March of 1974, he went so far as to make the following statement publicly: "In this room there is a representative of a country who is doing all he can to obstruct American policy. I want him to know that, if his country persists in this behavior, I'll crush it."[26] To achieve his foreign-policy goals using technological collaboration, Kissinger was not able to reinforce his policies with strong linkages; he protected them himself with his own aggressive form of hard power.

The line between what could be shared and what could be denied is not fixed; it changes over time and with circumstance, and it is necessarily contested. The decision as to what constitutes sensitive information or technology that can be shared without undermining US technological leadership is based on multiple, context-specific factors. It involves an assessment of the reliability of the partner seeking the technology and a clear idea of what its end use will be. It balances pressure from firms seeking export markets against government departments seeking security, and it calls for discrimination. Secretary for Defense Melvin Laird, responding to a request by the National Security Adviser to provide guidelines for collaboration with the French missile program, proposed that "sensitive information" be defined as "US information which, if released, could bring another party up to the US state of the art in areas where we believe we enjoy a technological advantage and where such technological advantage is a significant factor in our relative weapons capability or posture." Laird was quick to add that "of necessity the delineation of such sensitive areas is a matter of considered technical judgment."[27] First one had to determine whether the information would close the gap between the US and its partner. Then one had to decide whether closing that gap would be to the detriment of America's military assets. These prescriptions assumed American technological leadership, and were premised on the United States' commitment to ceaseless technological innovation that sought to keep it a generation ahead of everyone else.

We tend to forget that curbing proliferation also involves controlling the flow of "mundane" devices. "Everyday" devices such as high-speed computers also are dual-use devices. Allowing the French to use advanced computers in their military nuclear program "could save them time and money; it would enable them to perform more complex calculations and to produce more reliable and effective nuclear weapons."[28] Restrictions on the use of computers really irritated the French, who were ignoring their formal agreements not to use the computers they had bought from the US for military purposes. This problem so exercised the Nixon administration that Kissinger addressed the computer issue himself in a memo to the president.[29] He noted that machines equal in computing power to some advanced US models could be purchased from the FRG or from Japan. The US policy was needlessly harming relations between Washington and Paris, and was not applied to any other country anyway. Kissinger also noted that, since computer technology changed rapidly, the notion of an "advanced computer" needed frequent redefinition. This provided him with a solution that was formally spelled out in

National Security Decision Memorandum 103 of March 1971: "redefine 'advanced' computers' so that in practice some of the models currently falling under the restrictions on end-use will become available without any restrictions."[30] This meant allowing France unrestricted access to any computer whose power was rated below that of the IBM370/165.[31] More powerful computers could be exported, but doing so required a special letter assuring the vendor that it would not be used for military purposes. By 1975, the French, using that loophole, had bought four of Seymour Cray's CDC7600 computers, the most powerful machines on the market at the time. Although they were sold on the condition that they would not be used by weapons laboratories, everyone knew that they would be. In 1975 the French made a formal request to import a CDC7600 for military use. As one senior French official put it, they "would much prefer to have one that they could use with a clear conscience."[32] One could not stop the French from buying advanced computers, nor could one control what they did with them once they had bought them. Curbing proliferation by technological denial is a chimera in cases such as this.

I reminded readers earlier that a technological system was a loosely connected interdependent network of elements, some of which performed better than others; there was always room for improvement. The first round of requests for help from the French gives us some idea of how diverse such a system can be (and of the dual-use nature of many of its components). At issue was the performance of France's land-based and submarine-launched ballistic missiles. Secretary of Defense Melvin Laird explained to Kissinger that the French were not asking for help to develop missile systems; they had those already. "Rather," Laird said, "they are asking us to give them the benefit of our experience in solving some detailed—although not apparently catastrophic—problems, to save them the time and money involved in solving them on their own."[33] They were particularly concerned about the reliability, the quality control, and the maintenance programs of several components. Specific problem areas included solid-fuel propulsion systems (bonding, igniters, propellant characteristics), stress corrosion of nitrogen tanks, the longevity of gas bearings in gyroscopes, electrical connectors, missile pyrotechnic safety measures aboard submarines, and simulation techniques for use in hardening reentry vehicles.[34] NSDM103 of March 1971 allowed for assistance that would "not jeopardize the security of our own weapons programs or provide France with a distinct capability in such areas as guidance systems, missile accuracies, and re-entry vehicle hardening" (against the radiation produced by previous missiles fired at the target).[35] NSDM104 allowed for helping the French with information on "nuclear safety standards, safety devices and procedures, safety design, and command and control arrangements," and for using unclassified and classified data up to but not including restricted data.[36]

The longer the consultation continued, the more the French wanted—for example, they wanted to know how to successfully target multiple warheads on a single missile, and they wanted at least indirect information in the highly sensitive area of the design of warheads.[37] Kissinger was prepared to go as far as sharing knowledge about the physics of warheads, even though that was embargoed by Congress.

There are many ways to share information. As we saw in chapter 5, such sharing may involve handing over blueprints, giving presentations, engaging in informal discussions, and publishing papers. To help the French solve problems they were having with the booster trigger in their thermonuclear weapons, Kissinger proposed a method (subsequently dubbed Twenty Questions) to get around the strictures of the McMahon Act and the objections of Congress. He put it this way to Galley:

> We may not be able to give you information, but we can critique what you are doing. We can say "That's the wrong way." So there are many ways to give you information. ... It can be like a seminar; you can say you have three possibilities and we can tell you, "that's wrong; that's complicated," etc.[38]

This procedure, famously known as "negative guidance," was confirmed by the American scientists and engineers who were interviewed by Ullman. Of course there was no guarantee that it was followed to the letter. Several of Ullman's interview subjects confessed that they "had no confidence that anyone really knew what American scientists and engineers said to French colleagues over lunch and dinner once they had been given a basic license to talk." "As one put it," Ullman wrote, "'I found that the technicians just loved to share their triumphs; they had no inhibitions.'"[39]

Shared information can be put to many different uses. Even without strict reciprocity in the substance of the knowledge flowing back to the dominant partner, much can be acquired. For example, the American engineers who began sharing their knowledge of thermonuclear weapons with British allies in 1958 had little to learn from relative novices, but there were benefits all the same. An official explained this to the JCAE:

> For us, the merging of British information and know-how in a common fund with our own ... should provide two direct benefits. Since nuclear weapons programs have been carried out independently by each country we stand to gain from techniques developed by the British where they have solved the same problems which we faced by methods different from our own. For the same reason, where we find that British techniques developed separately are closely similar to those we have evolved, we can have added confidence in the evolution of our own program. Thus our weapons program should profit measurably through the stimulation which inevitably results from the cross-fertilization of ideas.[40]

Even the dominant partner can learn from an exchange of ideas with people who have tackled the same problems independently. Cutting-edge research is conducted at an uncertain epistemological frontier where micro-choices are made. There is no "one best way," and each choice has implications for the technological path that is chosen. As one scientist put it, "there may be half a dozen fundamentally different design approaches" to what you want to build. "Knowing that one of them worked out and did not run up against showstoppers can be pretty important."[41]

Knowledge circulates in interpersonal encounters under many guises and by a multitude of channels. Only by keeping those channels open can one benefit from the "cross-fertilization of ideas" and assess the quality and potential direction of another's work. After all, "technicians" have no inhibitions about sharing their "triumphs." Technological collaboration allows access to the programs of the other; denial closes down the dialogue. Collaboration may encourage proliferation, but it can also redirect or even retard it. Scientists and engineers who have access to the research sites and plans of their foreign colleagues can gain a pretty good idea of their level of competence, and can often infer from that and informal conversations what their future plans are. The proliferation risk of the gas centrifuge in the late 1960s was not contained only by classification, or by formal interstate arrangements enshrined in the US-UK agreements or in the tripartite international agreement that would help monitor Germany's nuclear program. It was also assessed in face-to-face encounters between researchers from the AEC and the UKAEA—people who operationalized those formal agreements and ensured that their terms were respected. Once the American weaponeers gained access to the French nuclear program, they could evaluate its strengths and weaknesses, could integrate it into their strategic planning, and could help in improving the existing missiles under development in the 1960s. American technological—and military—leadership was not challenged. In Kissinger's view this resulted in a far more secure situation than the prevailing alternative, which would have left the French to chart their own way with a highly vulnerable nuclear arsenal.

Technological collaboration is based on trust but can spawn mistrust. While always grateful for the help they got, the French were suspicious of American intentions and careful about what they revealed. They rejected offers from the US to test their weapons underground in Nevada for fear that they would have to disclose details about the designs and the performance of their warheads. Some suspected that the deadlock over the supply of the CDC7600 computer had been broken because the US hoped to gain access to the movements of France's nuclear submarines. One very senior official even claimed later that US scientists and engineers had deliberately fed the French false information during one of the exchanges on how to improve their nuclear warhead.[42]

Mutual distrust among the Six was given as a justification for self-inspection in Euratom, while lack of trust of Soviet inspectors was given as a justification for Euratom exercising self-inspection and sidelining the IAEA safeguards system. German industrialists were alarmed by the possibility for industrial espionage in Euratom. So as not give away industrial secrets, the AEC broke off close collaboration with the UKAEA just when researchers at Oak Ridge were about to develop their first production model of a gas centrifuge. Solly Zuckerman was convinced that commercial competition rather than security concerns motivated the AEC's demand for visual inspection of Britain's Mark I centrifuge.

Technological collaboration thus involves making difficult tradeoffs. It provides an avenue for knowledge circulation to mutual advantage, but it also opens research facilities to external scrutiny. It can keep lines of communication open until the knowledge concerned is deemed sensitive, whereupon collaborators can become competitors and sharing can give way to denial. It can de-escalate tension, but if abused by either party it can increase tension. From the point of view of a leading scientific and technological power such as the US, the access that technological collaboration provides can both reveal the level of the state of the art in other countries and provide new ideas to be exploited at home. It can help a government to formulate appropriate non-proliferation policies case by case. And it can channel research in partner countries along lines that cohere with American objectives.

Matthew Fuhrmann has recently argued that the circulation of sensitive nuclear technology and know-how made possible by programs such as Atoms for Peace has contributed to the spread of nuclear weapons.[43] This may be so. That technological collaboration entails proliferation risks cannot be denied. On the other hand, complete technological denial is not possible in practice. (Imposing international sanctions on a "rogue state" is an extreme exception.) It can violate Article IV of the Nuclear Non-Proliferation Treaty. It restricts the market for civilian nuclear materials. The challenge is to selectively combine denial with the sharing of sensitive technologies and scientific and managerial know-how; together they can reinforce each other to curb proliferation. In fact, for the senior US officials who are the subjects of this study, collaboration was always calibrated. By making fine distinctions in the technology that was shared in each case, they hoped to reap the strategic (and sometimes commercial) advantages of collaboration without encouraging proliferation. Broad generalizations in this field cannot capture the messiness of the decision-making process that leads states to collaborate with other states at the risk of proliferation, nor can they do justice to the sophisticated technological distinctions that are drawn to reap the benefits of collaboration while mitigating those risks as much as is practicable.

Roland Popp has recently suggested that it was only after India's test in 1974 that the "diplomatic emphasis" on curbing proliferation in the 1960s was "changed into a strategy of non-proliferation by denial" of sensitive nuclear technology.[44] The measures taken to stop states from getting nuclear weapons shifted from trying to get them to sign the Non-Proliferation Treaty to trying to stop them having access to pertinent technology. This periodization is true if one limits the analysis of non-proliferation strategies to international agreements and institutions, such as the Non-Proliferation Treaty (1968) and the formation of the Nuclear Suppliers Group (1974). It does not capture the efforts made by the US to control proliferation by sharing or denying technology from the dawn of the atomic age. To "see" and evaluate the role of technological collaboration, we need to probe behind diplomacy and situate the analysis at the "micro level" of transnational flows that circulate knowledge between scientists and engineers in different countries. These flows are authorized at the level of "high politics," of course, but their implementation often lies beyond the purview of the senior officials who agree to them.

Collaboration (and surveillance) at the level of interpersonal relations between expert scientists and engineers from different countries is difficult to track and falls outside the scope of most scholarly research on proliferation. Yet it is one of the pillars on which the security of a non-proliferation regime that exploits dual-use technologies rests.

There is no need to limit the policy debate on controlling proliferation to diplomatic and formal exchanges between senior government officials and between firms that supply nuclear equipment. The transnational circulation of scientists and engineers also plays a crucial role in defining and implementing non-proliferation policy. As such, it can serve as a useful weapon in a non-proliferation armory. John S. Foster, who left his post at the Lawrence Livermore Laboratory to work in the Department of Defense for eight years, was often present when Henry Kissinger discussed the kinds of technical support he could offer France's senior political and military representatives. It was Foster who was available to advise the White House on what knowledge could be safely shared with France and to translate Kissinger's diplomacy into calibrated face-to-face sharing of knowledge between scientists and engineers from the two countries.[45] Acting as vectors of their nation's interests abroad, and trained to draw lines between what can be shared and what must be denied, these experts can provide essential insights into the maturity of another country's nuclear program and into the risks of that country's progressing from a state of nuclear-weapons "pregnancy" to one of birth. The one-sided emphasis on diplomacy in (non)proliferation studies shrouds the core role of knowledge circulation in (de)stabilizing the international nuclear order. Transnational linkages between scientists and engineers who collaborate (or not) in circulating sensitive knowledge are constitutive of (non)proliferation regimes. As a matter of policy, and of scholarly reflection, it is also important to include them in the construction of a stable international system.

Notes

Introduction

1. Cited in John Newhouse, *De Gaulle and the Anglo-Saxons* (Viking, 1970), 165.

2. Marc Trachtenberg, *A Constructed Peace: The Making of the European Settlement 1945–1963* (Princeton University Press, 1999), chapter 6.

3. Memorandum of conversation, Dulles and Strauss, May 14, 1956, *FRUS* (*Foreign Relations of the United States*), 1955–1957, 4. 441.

4. For Kennedy and McNamara, see Newhouse, *De Gaulle and the Anglo-Saxons*, 157, 161, 188.

5. Source: https://en.wikipedia.org/wiki/Morgenthau_Plan

6. Trachtenberg, *A Constructed Peace*, 233.

7. George W. Ball, A Further Nuclear Offer to General de Gaulle? Memorandum to the President, August, 8, 1963, WCDA, Document 6.

8. Department of State Cable 5245, Message from President Kennedy to Prime Minister Macmillan, May 6, 1961, WCDA, Document 3.

9. Marc Trachtenberg, "The State of International History. Where We've Been, Where We Are, Where We Are Going," https://networks.h-net.org/node/28443/pages/33862/h-diplo-state-field-essays.

10. Department of State, United States Objectives and Programs for National Security, NSC-68, April 12, 1950, available at http://www.trumanlibrary.org/whistlestop/study_collections/coldwar/documents/pdf/10-1.pdf.

11. Michael J. Hogan, *A Cross of Iron: Harry S. Truman and the Origins of the National Security State* (Cambridge University Press, 1998), 15.

12. Hal Brands, "Non-Proliferation and the Dynamics of the Middle Cold War: The Superpowers, the MLF and the NPT," *Cold War History* 7:3 (2007): 389–423, at 397.

13. Joseph S. Nye Jr., "Soft Power," *Foreign Policy* 80:fall (1990): 153–172, at 167.

14. Joseph S. Nye Jr., *Soft Power: The Means to Success in World Politics* (Public Affairs, 2004).

15. The following works are representative: Michael Adas, *Dominance by Design: Technological Imperatives and America's Civilizing Mission* (Belknap, 2006); Ron E. Doel, "Science in Black: Scientific Intelligence and Scientific Practice in the Cold War Era," in *Cold War Science and the Transatlantic Circulation of Knowledge*, ed. Jeroen Van Dongen (Brill, 2015); Jacob Darwin Hamblin, "The Nuclearization of Iran in the Seventies," *Diplomatic History* 38:5 (2014): 1114–1135; Gabrielle Hecht, *Being Nuclear: Africans and The Global Uranium Trade* (MIT Press, 2012); Gabrielle Hecht, ed., *Entangled Geographies: Empire and Techno-politics in the Global Cold War* (MIT Press, 2011); Gabrielle Hecht and Paul N. Edwards, *The Technopolitics of the Cold War* (American Historical Association, 2007); Joy Rohde, *Armed with Expertise: The Militarization of American Social Research in the Cold War* (Cornell University Press, 2013); Mark Solovey and Hamilton Cravens, *Cold War Social Science: Knowledge Production, Liberal Democracy and Human Nature* (Palgrave Macmillan, 2014); Audra Wolfe, *Competing with the Soviets: Science, Technology, and the State in Cold War America* (Johns Hopkins University Press, 2014). Among diplomatic historians see, for example, David C. Engerman, "American Knowledge and Global Power," *Diplomatic History* 31:4 (2007): 599–621; Walter LaFeber, "Technology and US Foreign Relations," *Diplomatic History* 24:1 (2000): 1–19; Odd Arne Westad, "The New International History of the Cold War: Three (Possible) Paradigms," *Diplomatic History* 24:4 (2000): 551–565. On modernization in areas beyond Europe, see Matthew Connelly, *Fatal Misconception: The Struggle to Control World Population* (Belknap, 2006); Marcus Cueto, *Cold War Deadly Fevers: Malaria Eradication in Mexico, 1955–1975* (Johns Hopkins University Press, 2007); Nick Cullather, *The Hungry World: America's Cold War Battle Against Poverty in Asia* (Harvard University Press, 2010); Nick Cullather, "Miracles of Modernization: The Green Revolution and the Apotheosis of Technology," *Diplomatic History* 28:2 (2004): 227–254; David Engerman, Nils Gilman, Mark E. Haefele, and Michael E. Latham, eds., *Staging Growth: Modernization, Development, and the Global Cold War* (University of Massachusetts Press, 2003); Susan Greenhalgh, "The Social Construction of Population Science: An Intellectual, Institutional and Political History of Twentieth-Century Demography," *Comparative Studies in Society and History* 38:1 (1996) 26–66; Michael E. Latham, *The Right Kind of Revolution: Modernization, Development and US Foreign Policy from the Cold War to the Present* (Cornell University Press, 2010).

16. Cristina Klein, *Cold War Orientalism: Asia in the Middlebrow Imagination, 1945–1961* (University of California Press, 2003), 16.

17. John Krige, Angelina Long Callahan, and Ashok Maharaj, *NASA in the World: Fifty Years of International Collaboration in Space* (Palgrave Macmillan, 2013), chapter 7; Erez Manela, "A Pox on Your Narrative: Writing Disease Control into Cold War History," *Diplomatic History* 34:2 (2010): 199–323.

18. Scott D. Sagan, "Why Do States Build Nuclear Weapons? Three Models in Search of a Bomb," *International Security* 21:3 (1996–97): 54–86.

19. Angela Creager, *Life Atomic: A History of Radioisotopes in Science and Medicine* (University of Chicago Press, 2013).

20. Gabrielle Hecht, *The Radiance of France: Nuclear Power and National Identity After World War II* (MIT Press, 1998); Martin Kohlrausch and Helmuth Trischler, *Building Europe on Expertise: Innovators, Organizers, Networkers* (Palgrave Macmillan, 2014); Walter A. McDougall, "Space Age Europe: Gaullism, Euro-Gaullism and the American Dilemma," *Technology and Culture* 26:2 (1985): 179–203; Helmuth Trischler and Hans Weinberger, "Engineering Europe: Big Technologies and Military Systems in the Making of Twentieth Century Europe," *History and Technology* 25:1 (2005): 49–83.

21. Donald Hornig, Visit to Germany and Opportunities for Cooperation, memo for the President, January 31, 1966, LBJL, Papers of D. Hornig, Director, OST, box 3, folder Hornig Chrono File, Jan-Feb 1966.

22. John Krige, "Atoms for Peace, Scientific Internationalism and Scientific Intelligence," in *Global Power Knowledge. Science and Technology in International Affairs*, ed. John Krige and Kai-Henrik Barth (*Osiris* 21, 2006); Martin J. Medhurst, "Atoms for Peace and Nuclear Hegemony: The Rhetorical Structure of a Cold War Campaign," *Armed Forces and Society* 23 (1997): 571–593.

23. Krige et al., *NASA in the World*, chapter 1.

24. For a comprehensive bibliography, see John Krige and Helke Rausch, eds., *American Foundations and the Coproduction of World Order in the Twentieth Century* (Vandenhoeck & Ruprecht, 2012).

25. Daniel J. Kevles, "Principles and Politics in Federal R&D Policy, 1945–1990. An Appreciation of the Bush Report," in Vannevar Bush: Science—The Endless Frontier, National Science Foundation report NSF90–8, 1990.

26. To cite three classic texts: Walter A. McDougall, *The Heavens and the Earth: A Political History of the Space Age* (Johns Hopkins University Press, 1997); Alex Roland, *The Military-Industrial Complex* (AHA-SHOT Publications, 2001); Ruth Oldenziel and Karin Zachmann, eds., *Cold War Kitchen: Americanization, Technology, and European Users* (MIT Press, 2011).

27. Aaron L. Friedberg, *In the Shadow of the Garrison State: America's Anti-Statism and its Cold War Grand Strategy* (Princeton University Press, 2000), 297.

28. Truman's inaugural address, January 20, 1949, available at http://www.trumanlibrary.org/whistlestop/50yr_archive/inagural20jan1949.htm.

29. I make much of this in "Diplomacy (Post 1945), Science and Technology and" in *Oxford Encyclopedia of the History of American Science, Medicine and Technology*, volume I, ed. Hugh R. Slotten (Oxford University Press, 2014).

30. Stuart A. Butler, "The Struggle for Power: Britain and Euratom 1955–1963," *International History Review* 36:2 (2014): 324–341.

31. John Krige, *American Hegemony and the Postwar Reconstruction of Science in Europe* (MIT Press, 2006).

32. Charles Weiner, "A New Site for the Seminar: The Refugees and American Physics in the Thirties," in *The Intellectual Migration: Europe and America, 1930–1960*, ed. D. Fleming and B. Bailyn (Belknap, 1969).

33. Cited in Georges-Henri Soutou, "La politique nucléaire de Pierre Mendès France," in *La France et l'atome: Études d'histoire nucléaire*, ed. Maurice Vaîsse (Bruylant, 1994).

34. Alan S. Milward, *The United Kingdom and the European Community*, volume I: *The Rise and Fall of National Strategy 1945–1963* (Routledge, 2013), 225.

35. Jacques Blamont, "Origines et principes de la politique spatiale de la France," Rayonnement du CNRS 48, June 2008: 24–32.

36. David Edgerton, *Warfare State: Britain 1920–1979* (Cambridge University Press, 2005).

37. Simone Turchetti, "A Most Active Customer: How the US Administration Helped the Italian Atomic Energy Project to 'De-Develop,'" *Historical Studies in the Natural Sciences* 44:5 (2014): 470–502.

38. Hamblin, "The Nuclearization of Iran in the Seventies," at 1119.

39. William Burr, "A Scheme of 'Control': The United States and the Origins of the Nuclear Suppliers' Group, 1974–1976," *International History Review* 36:2 (2014): 252–276.

40. Hecht, *Being Nuclear*.

Chapter 1

1. For an overview of the current argument with less emphasis on the diplomatic angle, see John Krige, "The Peaceful Atom as Political Weapon: Euratom and American Foreign Policy in the Late 1950s," *Historical Studies in the Natural Sciences* 38:1 (2008): 5–44.

2. Euratom usually receives little more than a passing reference. The following works are exceptions: Michel Dumoulin, Pierre Guillen, and Maurice Vaïsse, eds., *L'Énergie nucléaire en Europe: Des origins à Euratom* (Peter Lang, 1994); Jonathan D. Helmreich, "The United States and the Formation of Euratom," *Diplomatic History* 15:3 (1991): 387–410; Pascaline Winand, *Eisenhower, Kennedy, and the United States of Europe* (St. Martin's Press, 1993), chapter 4; Gunnar Skogmar, *The United States and the Nuclear Dimension of European Integration* (Palgrave Macmillan 2004). Marc Trachtenberg deals briefly with the topic in *A Constructed Peace*. For analyses from a UK point of view, see Butler, "The Struggle for Power"; M. Elli, *Politica estera ed ingegneria nucleare: I rapporti del Regno Unito con l'Euratom, 1957–1963* (Unicopli, 2007); Alan S. Milward, *The United Kingdom and the European Community*, volume 1: *The Rise and Fall of a National Strategy* (Routledge, 2002), chapter 8; Mervyn O'Driscoll, "Missing the Nuclear Boat: British Policy and French Nuclear Ambitions during the Euratom Foundation Negotiations, 1955–6," *Diplomacy and Statecraft* 1 (1998): 135–162.

3. Geir Lundestad, *"Empire" by Invitation: The United States and European Integration, 1945–1997* (Oxford University Press, 1998), 134.

4. A memo by Schaetzel ("Meeting with Mr. Rubottom Regarding Euratom—Tuesday February 26," February 25, 1957) states that "the fourth nation nuclear

weapon problem is in the immediate future largely a French and German problem." NARA, RG 59, box 364, folder 19.8, Regional Program, Euratom, General, Jan-Feb 1957, part 2 of 2.

5. Klaus Schwabe, "The United States and European Integration," in *Western Europe and Germany: The Beginnings of European Integration, 1945–1960*, ed. Clemens Wurm (Berg, 1995), 129.

6. Skogmar, *The United States and the Nuclear Dimension of European Integration*, 253, 257.

7. Trachtenberg, *A Constructed Peace*, 150, 151.

8. Memorandum of a Conversation, Department of State, May 14, 1956, *FRUS*, 1955–1957, 4: 441.

9. Grégoire Mallard and Andrew Lakoff, "How Claims to Know the Future Are Used to Understand the Present," in *Social Science in the Making*, ed. Charles Camic, Michèle Lamont, and Neil Gross (University of Chicago Press, 2011), at 344.

10. Grégoire Mallard, *Fallout: Nuclear Diplomacy in an Age of Global Fracture* (University of Chicago Press, 2014), 153.

11. Jack M. Holl, "The Peaceful Atom: Lore and Myth," in *Atoms for Peace: An Analysis after Thirty Years*, eds. Joseph F. Pilat, Robert E. Pendley, and Charles K. Ebinger (Westview), 154.

12. Quoted in Patrick J. McGrath, *Scientists, Business, and the State, 1890–1960* (University of North Carolina Press, 2002), 177.

13. For accounts of the Baruch Plan from very different points of view, see Shane J. Maddock, *Nuclear Apartheid: The Quest for American Supremacy From World War II to the Present* (University of North Carolina Press, 2010), chapter 3; Henry D. Sokolski, *Best of Intentions: America's Campaign Against Strategic Weapons Proliferation* (Praeger, 2001), chapter 2.

14. On Operation Candor, see Richard G. Hewlett and Jack M. Holl, *Atoms for Peace and War, 1953–1961: Eisenhower and the Atomic Energy Commission* (University of California Press, 1989), 42–60. See also Ira Chernus, *Eisenhower's Atoms for Peace* (Texas A&M University Press, 2002).

15. *FRUS*, 1952–1954, 2.2:1526.

16. On Strauss, Oppenheimer and Operation Candor, see Hewlett and Holl, *Atoms for Peace and War*, 50–56; on the effects of the Soviet test, see ibid., 57–61. I thank Alex Wellerstein for helpful insights here.

17. *FRUS*, 1952–1954, 2-2: 1527.

18. Ibid.: 1213.

19. Hansard, HC Debates, volume 522, December 17, 1953, c. 585.

20. Source: http://voicesofdemocracy.umd.edu/eisenhower-atoms-for-peace-speech-text/

21. Memo by Cutler, September 10, 1953, *FRUS*, 1952–1954, II-2, 1213.

22. Lawrence Scheinman, *The International Atomic Energy Agency and World Order* (Resources for the Future, 1987), 59.

23. Stephen Twigge, "The Third Nuclear Power. Britain and the Beginnings of the IAEA," paper presented at conference on United Atoms in a Divided World, Vienna, 2012.

24. Cited by Henry Sokolski in "The Arms Control Connection," in *Atoms for Peace*, ed. Pilat et al., 43.

25. Greg Herken, "'A Most Deadly Illusion': The Atomic Secret and American Nuclear Weapons Policy, 1945–1950," *Pacific Historical Review* 49:1 (1980): 51–76.

26. Howard Morland, "Born Secret," *Cardozo Law Review* 26:4 (2005): 1401–1408; Arvin S. Quist, Security Classification of Information, volume I: Introduction, History and Adversary Impacts, Report ORCA-12, Oak Ridge Classification Associates, LLC, September 2002, chapter 4.

27. This paragraph has been crafted with major inputs from Alex Wellerstein.

28. Dwight D. Eisenhower, address at centennial commencement, Pennsylvania State University, June 11, 1955 (http://www.presidency.ucsb.edu/ws/index.php?pid10256).

29. United States Delegation to the International Conference on the Peaceful Uses of Atomic Energy, Technical Exhibition of the United States of America, volume 2, appendix V—Research Reactor Exhibit at Geneva, Special Collections Division, Georgetown University Library, Washington, 314–315.

30. This paragraph briefly summarizes the following works: Krige, "Atoms for Peace," in Krige and Barth (*Osiris* 21, 2006); John Krige, "Techno-Utopian Dreams, Techno-Political Realities. The Education of Desire for the Peaceful Atom," in *Utopia/Dystopia: Conditions of Historical Possibility*, ed. Michael D. Gordin, Helen Tilley, and Gyan Prakash (Princeton University Press, 2010).

31. In *Proceedings of the International Conference on the Peaceful Uses of Atomic Energy, Held in Geneva, 8 August–20 August 1955*, volume 16, *Record of the Conference* (United Nations, 1956), 31, 33. For an analysis of Bhabha's opening speech, see Itty Abraham, *The Making of the Indian Atomic Bomb* (Zed, 1998), 98–102.

32. Geminus, "It Seems to Me," *New Scientist*, September 4, 1955, 742.

33. Memo, Barnett to Merchant, August 19, 1955, *FRUS*, 1955–57, 4: 326.

34. Kenneth Osgood, *Total Cold War: Eisenhower's Secret Propaganda Battle at Home and Abroad* (University Press of Kansas, 2006), 155–156.

35. Jackson, letter dated November 10, 1953, quoted by Spencer R. Weart in *Nuclear Fear: A History of Images* (Harvard University Press, 1998).

36. Hewlett and Holl, *Atoms for Peace and War*, appendix 6.

37. William Walker and Måns Lönnroth, "Arms Control or Anarchy," in *Atoms for Peace*, ed. Pilat et al., 171.

38. Osgood, *Total Cold War*, 176.

39. I discuss the case study at length in "The Peaceful Atom as Political Weapon." A far more extensive list of references can be found there.

40. Memorandum of a conversation with Albert Coppe, Department of State, April 20, 1955, *FRUS*, 1955–57, 4: 286.

41. Dulles to Mayer, June 7, 1955, *FRUS*, 1955–57, 4: 293.

42. Discussion with Ludwig Erhard, June 7, 1955, *FRUS*, 1955–1957, 4: 292.

43. *FRUS*, 1955–57, 4: 349.

44. A Combined Development Agency (originally Trust), set up in 1946, continued the wartime agreement whereby the United States, and after it Britain, had access to all the uranium mined in Katanga in the Belgian Congo. In return, Belgium was given fissile material and classified information on power reactors not generally made available to foreign governments. On the sources of uranium and their output, see Luc Gillon, "L'approvisionnement en uranium," in *L'Énergie nucléaire en Europe*, ed. Dumoulin, Guillen, and Vaïsse. On the colonial history of uranium mining, see Hecht, *Being Nuclear*.

45. Merchant to Dulles, July 1, 1955, *FRUS*, 1955–57, 4: 305.

46. Smith to Dulles, July 1, 1955, *FRUS*, 1955–57, 4: 306.

47. Telegram to Geneva, July 21, 1955, *FRUS*, 1955–57, 4: 322, note 3.

48. Memorandum of conversation, Department of State, July 15, 1955, *FRUS*, 1955–57, 4: 313–318.

49. Memorandum, Barnett to Merchant, August 19, 1955, *FRUS*, 1955–57, 4: 326.

50. Memo, Martin to Timmons, November 10, 1955, *FRUS*, 1955–57, 4: 347.

51. Winand, *Eisenhower, Kennedy, and the United States of Europe*, 51.

52. Vultee to Van Dyke, November 9, 1955, copied to Barnett, Schaetzel and others, November 29, 1955, NARA, RG59 box 363, folder Regional Programs General Sep-Oct 1955 thru May–Sep 1956.

53. The terms are used by Smith to characterize Barnett's memo in his reply to Merchant, December 6, 1955, *FRUS*, 1955–57, 4: 360.

54. Memorandum drafted by Barnett in the European Office of Regional Affairs, December 6, 1955, *FRUS*, 1955–57, 4: 355–360.

55. Robinson to Bowie, December 27, 1955, *FRUS*, ,1955–57, 4: 378–379.

56. Lawrence Scheinman, *Atomic Energy Policy in France Under the Fourth Republic* (Princeton University Press, 1965), 134–138.

57. Pierre Guillen, "La France et la négociation du traité d'Euratom," in *L'Énergie nucléaire en Europe*, ed. Dumoulin et al.

58. L. Nicolaidis, R. Ockrent, and W. Harpham, *Possibilities of Action in the Field of Nuclear Energy* (Organization for European Economic Cooperation, 1956).

59. Mollet's view on this is cited at length on p. 137 of Scheinman, *Atomic Energy Policy in France*. See also Bertrand Goldschmidt, *The Atomic Complex: A World-*

wide Political History of Nuclear Energy (American Nuclear Society, 1982), 289–302.

60. Memorandum of conversation, November 25, 1955, NARA, RG59, box 363, folder 19.8 Regional Program Euratom General Nov—Dec 1955,.

61. Quoted by Scheinman, *Atomic Energy in France*, 137–138.

62. Note for the Record, written by William Strath (UKAEA), on a meeting between Spaak and the Lord President, Sir Edwin Plowden, the first chairman of the UKAEA and Strath, London, November 17, 1955, UKNA, FO371/116054.

63. Wolfram Kaiser, *Using Europe, Abusing the Europeans: Britain and European Integration, 1945–63* (St. Martin's Press, 1996), 48.

64. Quoted by Kaiser, *Using Europe*, 49. The memo in question is dated November 11, 1955.

65. Peter Nichols, quoted by John W. Young, *Britain and European Unity, 1945–1992* (St. Martin's Press, 1993), 47.

66. N. Piers Ludlow, "European Integration and the Cold War," in *The Cambridge History of the Cold War*, volume II: *Crisis and Détente*, ed. Melvyn P. Leffler and Odd Arne Westad (Cambridge University Press, 2010), 188.

67. Memo, A. A. Jukes to W. Strath, with attached Policy on International Cooperation, November 1, 1955, UKNA, AB16/1727.

68. See note 44.

69. For a report on the event, see "1956: Queen Switches on Nuclear Power" at http://news.bbc.co.uk/onthisday/hi/dates/stories/october/17/newsid_3147000/3147145.stm. For technical details, see "Calder Hall Power Station, No. I," *The Engineer*, October 5, 1956, 464–468.

70. Telegram from the Foreign Office to British embassies in the Six, and to Washington, November 15, 1955, UKNA, FO371/16054.

71. Letter, Harold Caccia to (illegible) in Geneva, November 14, 1955, UKNA, FO371/116054.

72. Michael L. Hoffman, "Britain Exploits Atomic Market," *New York Times*, August 8, 1955.

73. "OEEC and Nuclear Energy," Report on a meeting of the President of the Board of Trade with Plowden and others on March 8,1956, UKNA AB16/1958.

74. Memo, Strath to UKAEA Chairman (Plowden), March 7, 1956, UKNA SB16/1958.

75. Memo, Strath to Plowden, February 26, 1957, UKNA, AB16/3711.

76. Telegram from Foreign Office to Washington, November 15, 1955, and aide-mémoire, November 17, 1955, UKNA, FO371/16054.

77. Letter, Dulles to Macmillan, December 10, 1955, *FRUS*, 1955–57, 4: 362.

78. Message from Dulles to Eisenhower reporting on his talks with Spaak, Monnet, and Brentano, December 17, 1955, *FRUS*, 1955–57, 4: 369. See also memorandum of a Conversation, Monnet, Dulles, Bowie, December 17, 1955, *FRUS*, 1955–57, 4: 367.

79. Telegram from US ambassador in The Hague to Department of State, December 21, 1955, *FRUS*, 1955–57, 4: 377.

80. Kaiser, *Using Europe*, 50.

81. Telegram from Conant to Department of State, November 4, 1955, *FRUS*, 1955–57, 4: 344–348.

82. Extracts from a report published in *The Economist*, NARA, RG59 box 363, folder 19.8 Regional Program Euratom General Jan-Feb 1956, part 1 of 2, 5 Jan 1956.

83. Letter, Hoyer-Miller, UK embassy Bonn, to Caccia, February 8, 1956, UKNA FO371/121950.

84. Memo, Dulles to Eisenhower, January 9, 1956, *FRUS*, 1955–57, 4: 389–390.

85. Letter, Dulles to the Foreign Office, December 13, 1955, UKNA, FO371/116056.

86. Letter, Dulles to Macmillan, December 10, 1955.

87. Telegram, Dulles to Department of State, December 17, 1955, *FRUS*, 1955–57, 4: 370.

88. Memo, Merchant and Smith to Dulles, January 19, 1956, *FRUS*, 1955–57, 4: 391.

89. Memo of a Conversation on Atomic Energy and European Integration, January 25, 1956, *FRUS*, 1955–57, 4: 390–399.

90. Memo of a Conversation on Atomic Energy and European Integration, January 25, 1956, *FRUS*, 1955–57, 4: 390–399.

91. Letter, Strauss to Dulles, April 13, 1956, *FRUS*, 1955–57, 4: 423 and attachment.

92. Telegram from Secretary of State to embassy in Belgium, May 24, 1956, *FRUS*, 1955–57, 4: 442–444.

93. All quotations in this paragraph are from the telegram cited in the preceding note.

94. Conant to Merchant, February 19, 1956, NARA, RG59, box 363, folder 19.8, Regional Program Euratom General, Jan-Feb 1956, part 1 of 2.

95. Memorandum of conversation, Department of State, May 14, 1956, *FRUS*, 1955–57, 4: 438–439.

96. Memorandum of a Conversation, Department of State, May 14, 1956, *FRUS*, 1955–57, 4: 435–436.

97. Memorandum of a Conversation between Timmons and Strauss, May 14, 1956, *FRUS*, 1955–57, 4: 437–438.

98. Memo, American embassy in Bonn to Department of State, "Conversation with Minister Strauss on Euratom atomic energy integration," February 6, 1956, NARA, RG59, box 363, folder 19.8, Regional Program Euratom General Jan-Feb, 1956, part 1 of 2.

99. Memorandum of conversation, May 14, 1956, *FRUS*, 1955–57, 4: 435–436. See also letter Conant to Merchant, February 10, 1956, NARA, RG59, box 363, folder 19.8, Regional Program Euratom General Jan-Feb, 1956, part 2 of 2.

100. This formulation is from a commissioned report by Klaus E. Knorr: Euratom and American Policy: A Conference Report (Center for International Studies, Princeton University, 1956), available in NARA, RG59, box 363, folder 19.8 Regional Programs Euratom General May–September, 1956, part 1 of 2.

101. Memo from American embassy in Bonn to Department of State, February 6, 1956, n.98.

102. Letter, Hoyer Millar to Caccia, November 21, 1955, UKNA, FO371/116056.

103. Hans-Peter Schwartz, *The Statesman, 1952–1967* (Berghahn Books, 1997), 124.

104. Memo from American embassy in Bonn to Department of State, February 6, 1956, n.98.

105. Hewlett and Holl, *Atoms for Peace and War*, 113.

106. Memo of conversation, Department of State, May 14, 1956, *FRUS*, 1955–57, 4: 435–436.

107. Newhouse, *De Gaulle and the Anglo-Saxons*, 60.

108. In fact Heisenberg had no interest in a weapons program, and he and the German scientific community were not even consulted over the terms of the Euratom agreements. See Cathryn Carson, *Heisenberg in the Atomic Age: Science and the Public Sphere* (Cambridge University Press, 2010), 244. See also Michael Eckert, "Primacy Doomed to Failure: Heisenberg's Role as Scientific Adviser for Nuclear Policy in the FRG," *Historical Studies in the Physical and Biological Sciences* 21:1 (1990): 29–58.

109. Letter, Conant to Merchant, February 10, 1956, n. 99.

110. NSC162/2, Basic National Security Policy, Washington, October 30, 1953, Clause 39b.

111. By November of 1956, some German scientists, including Heisenberg, were increasingly concerned that both Adenauer and Strauss (now Minister of Defense) favored producing nuclear weapons in Germany. A letter that anticipated the so-called Göttingen Manifesto, publicly released on April 12, 1957, and signed by eighteen leading members of the German physics community, informed Strauss that "none of us would be prepared to participate in German production of nuclear weapons." Carson, *Heisenberg in the Atomic Age*, 321.

112. John Foster Dulles, Memorandum of a Conversation, Washington, February 6, 1956, *FRUS*, 1955–1957, 4: 406–407.

113. Trachtenberg, *A Constructed Peace*, 150.

114. Memo, Timmons (Director of Office of Regional Affairs), "Conversation with Franz Josef Strauss," NARA, RG59, box 363, folder 19.8, Regional Programs Euratom General May–September, 1956, part 1 of 2. This text is not included in the version of this conversation published in *FRUS*, 1955–57, 4: 437–438.

115. Memo of a Conversation, May 14, 1956, *FRUS*, 1955–57, 4: 438–441.

116. Georges-Henri Soutou, "La politique nucléaire de Pierre Mendès France," in *La France et l'atome: Études d'histoire nucléaire*, ed. Maurice Vaïsse (Bruylant, 1994).

117. Ibid., 96.

118. Guillen, "La France et la négociation du traité d'Euratom."

119. Telegram from Dillon to Department of State, June 10, 1955, *FRUS*, 1955–57, 4: 293–295.

120. On the ore, see Hecht, *Being Nuclear*. On France's civilian program see also Hecht, *The Radiance of France: Nuclear Power and National Identity After World War II* (MIT Press, 2009).

121. Guillen, "La France et la négociation du traité d'Euratom."

122. Scheinman, *Atomic Energy Policy in France*, 164.

123. "Current Status of Euratom," attached to letter from Murphy to Strauss, August 7, 1956, *FRUS*, 1955–57, 4: 457–460.

124. Internal Memo, Schaetzel to Smith and Timmons, August 2, 1956, NARA, RG59, box 364, folder 19.8, Regional Program Euratom General, May–September 1956, part 2 of 2.

Chapter 2

1. This account draws heavily on pages 177–183 of William I. Hitchcock's *The Struggle for Europe: The Turbulent History of a Divided Continent, 1945 to the Present* (Anchor Books, 2004).

2. Hitchcock, *The Struggle for Europe*, 181.

3. Memorandum of conversation between Spaak and senior officials, including Dulles and Strauss, February 8, 1957, *FRUS*, 1955–1957, 4: 519–523.

4. (Topic which may be raised by Wise Men or Spaak) Euratom and Nuclear Weapons, undated, unsigned, NARA, RG 59, box 362, folder 19.8s, Euratom Visits. 4. Wisemen, 1957.

5. Current Status of Euratom Negotiations, December 3, 1956, *FRUS*, 1955–1957, 4: 491–495.

6. Letter, Strauss to Dulles, *FRUS*, 1955–1957, 4: 537–542.

7. Footnote 2 to Telegram Dulles to embassy in Belgium, January 24, 1957, *FRUS*, 1955–1957, 4: 505–506.

8. Telegram from Dulles to US embassy in Belgium, January 24, 1957, *FRUS*, 1955–1957, 4: 505–506.

9. Current Status of Euratom Negotiations, December 3, 1956, *FRUS*, 1955–1957, 4: 491–495.

10. Aide-Memoire from MM. Armand, Etzel, and Giordani to the Secretary of State and the Chairman of the Atomic Energy Commission, February 4, 1957,

NARA, RG 59, box 362, folder 19.8s, Euratom Visits. 4. Wisemen, 1957. See also Memorandum of conversation, Department of State, Washington, February 4, 1957, *FRUS*, 1955–1957, 4: 512–515.

11. Memorandum of Conference with the President, ... February 6, 1957 *FRUS*, 1955–1957, 4: 516–518.

12. Excerpt from Department of State Policy Report Current Foreign Relations Issue No.7, February 13, 1957, NARA, RG 59, box 362, folder 19.8s, Euratom Visits. 4. Wisemen, 1957.

13. Letter, Schaetzel to Sprouse, March 8, 1957, NARA, RG 59, box 362, folder 19.8s, Euratom Visits. 4. Wisemen, 1957. The full report on the meeting with industrialists is also available in this folder.

14. Armand during a meeting with the AEC, February 5, 1957, NARA, RG 59, box 362, folder 19.8s, Euratom Visits. 4. Wisemen, 1957.

15. "The Wise Men Report," Atomic Industrial Forum Inc., Forum Memo, June 1957, NARA, RG 59, box 365, folder Regional Program Euratom General June–December 1957, part 1 of 3.

16. Bob Schaetzel, Briefing Memorandum for Meeting with the Joint Committee on Atomic Energy Regarding Euratom, Tab I-A, March 28, 1957, NARA, RG 59, box 364, folder Regional Program Euratom General March-April 1957, part 1 of 2.

17. Schaetzel to Smith, March 26, 1957, Briefing Memorandum For Meeting with Joint Committee on Atomic Energy Regarding Euratom, NARA, RG 59, box 364, folder Regional Program Euratom General March-April 1957, part 1 of 2.

18. Memo prepared in the Office of Regional Affairs, drafted by Barnett, December 6, 1955, *FRUS*, 1955–1957, 4: 355–360.

19. The general narrative is from R. L. Yost to Schaetzel, European Isotopic Separation Plant, July 9, 1957, NARA, RG 59, box 365, folder19.8, Regional Program Euratom General June–December 1957, part 1 of 3.

20. Declaration of the Foreign Ministers of the CSC Countries Adopted February 4, 1957, NARA, RG 59, box 364, folder 19.8, Regional Program Euratom General May 1957.

21. Report of the Study Panel, May 17, 1957, NARA, RG 59, box 364, folder 19.8, Regional Program Euratom General May 1957.

22. Background Paper, Topic which may be Raised by Wise Men or Spaak, Isotope Separation, NARA, RG 59, box 362, folder 19.8s, Euratom Visits, 4. Wisemen 1957.

23. Schaetzel to Butterworth, July 26, 1957; memo, Goldberg to Schaetzel, "Draft on European Tour," August 1, 1957; Sprouse to Schaetzel, August 9, 1957, all in NARA, RG 59, box 362, folder 19.8s, Euratom Visits. 2. Fifty European industrialists.

24. Schaetzel, "Department Tactics with Respect to Western European Power Bilateral Agreements," January 25, 1957, NARA, RG 59, box 356, folder 19.06, Regional Programs General 1960 thru ...; folder19.08j Documents 1957.

25. R.B. Freund, Conversation with Signor Roberto Ducci, January 17, 1957, NARA, RG 59, box 364, folder 19.08, Regional Programs General May–September 1956thru … May 1957, part 2 of 2.

26. Letter, Strauss to the Undersecretary of State (Hoover), December 19, 1956, *FRUS*, 1955–1957, 4: 495–497; memo of conversation with Mr. Jean Monnet, January 18, 1957, NARA, RG 59, box 364, folder 19.08, Regional Programs General May–September 1956 thru … May 1957, part 2 of 2.

27. Letter, Smith to Strauss, February 6, 1957, NARA, RG 59, box 356, folder 19.06, Regional Program General 1960 thru … ; folder19.08j, Documents 1957. See also letter from Dulles to Strauss, January 29, 1957, *FRUS*, 1955–1957, 4: 510–511; memo of conversation, Cleveland with Noel on the bilateral with France, February 27, 1957, NARA, RG 59, box 364, folder 19.8, Regional Program Euratom General Jan-Feb 1957, part 2 of 2.

28. Schaetzel, "Next Steps Regarding Euratom," April 22, 1957, NARA, RG 59, box 364, folder 19.8, Regional Program Euratom General March-April 1957, part 2 of 2.

29. Telegram from Dulles to US embassy in Germany, May 21, 1957, *FRUS*, 1955–1957, 4: 553–555.

30. Memo, Schaetzel to Smith and Jones, "Next steps regarding Euratom," April 22, 1957, NARA, RG 59, box 364, folder 19.8, Regional Program Euratom General Mar-Apr 1057, part 2 of 2. The FRG eventually got four research reactors, two small demonstration power reactors, and one 100-megawatt reactor with this help. See Michael Eckert, "Kernenergie und Westintegration: Die Zähmung des westdeutschen Nuklearnazionalizmus," in *Vom Marshallplan zur EWG: Die Engliederung der Bundesrepublik Deutschland in die westliche Welt*, ed. Ludolf Herbst, Werner Bührer, and Hanna Sowade (Oldenbourg, 1990), 313.

31. Schaetzel to Butterworth, July 26, 1957, NARA, RG 59, box 362, folder 10.8s, Euratom Visits 2. Fifty European Industrialists, 1957.

32. Letter, Barnett to Butterworth, October 28, 1957, NARA, RG 59, box 365, folder 19.8, Regional Program Euratom General, June-December 1957, part 2 of 3.

33. *Council of Europe News*, January 1, 1958, NARA, RG 59, box 365, folder 19.8, Regional Program Euratom General 1958, part 1 of 2.

34. Alan Milward, *The Rise and Fall of a National Strategy: The UK and The European Community*, volume 1 (Routledge, 2013), 226.

35. For this paragraph, see memo by Amasa S. Bishop, Discussion of a Joint Cooperative Program with Euratom, November 27, 1957, *FRUS*, 1955–1957, 4: 566–569.

36. Letter, Barnett (US embassy, The Hague) to Schaetzel, September 10, 1956, NARA, RG59, box 364, folder 19.8, Regional Program Euratom General May–September 1956, part 2 of 2.

37. Schaetzel, Factors Affecting a US-Euratom Nuclear Power Demonstration Program, attached to Memorandum for the Secretary, January 7, 1958, NARA, RG 59 box 361, folder 19.08r7, Regional Programs Joint Program 1962 thru …

1958 [2 of 2]. See also Excerpt of AEC Draft of NSC Policy Paper, attached to letter Schaetzel to Butterworth, September 19, 1957, NARA, RG 59, box 365 folder 19.8, Regional Program Euratom General June–December 1957, part 1 of 3.

38. James R. Killian, *Sputniks, Scientists and Eisenhower*, 7.

39. John Krige, "NATO and the Strengthening of Western Science in the Post-Sputnik Era," *Minerva* 38 (2000): 81–108, at 93.

40. Dwight D. Eisenhower, State of the Union Address, January 9, 1958 (http://www2.hn.psu.edu/faculty/jmanis/poldocs/uspressu/SUaddressDEisenhower.pdf).

41. Opening Statement Before Joint Committee on Atomic Energy by Mr. Dillon with Respect to the US-Euratom Joint Program, NARA, RG 59, box 360, folder R08r4, Regional Programs Fact Sheets 1959–1960 thru … Joint Program 1961.

42. The description that follows is from that provided by AEC Chairman Lewis Strauss attached to the final agreement, NARA, RG 59, box 359, folder 19.08r, 1. Agreement for Peaceful Uses, part 3 of 3.

43. The president's letter and the international agreement are in NARA, RG 59, folder 19.8r, 1. Agreement for Peaceful Uses, 3 of 3.

44. Schaetzel in Study Group Report, Peaceful Uses of Atomic Energy, Council of Foreign Relations, May 7, 1958, NARA, RG 59, box 367, folder 19.8j, Regional Programs Documents, 1958.

45. Dispatch from UK embassy, 5 January 1959, NARA, RG 59, box 360, folder 19.8r, 5. Financial Arrangements Euratom.

46. Memo, UK embassy, 25 Nov 1958, NARA, RG 59 box 359 folder 19.8r, Agreement for Peaceful Uses, part 3 of 3.

47. Memo, Werner to Wells, British Protest of US-Euratom Reactor Fuel Guarantees, December 11, 1958, NARA, RG 59, box 360, folder 19.08r4, Regional Programs Fact Sheets 1959 to 1960, thru … Joint Program 1961.

48. Draft memo referring to UK memo of 25 Nov 1958, 14 Jan 1959. NARA, RG59, box 357, folder 19.8m.

49. Baylis, "Exchanging Nuclear Secrets," at 48.

50. Schaetzel to Isenbergh, 26 March 1957, NARA, RG 59 box 364 folder Regional Programs Euratom General Mar—Apr 1957, part 1 of 2.

51. Schaetzel to Isenbergh, 26 March 1957.

52. Jardrey to Barbour, February 11, 1958, NARA, RG 59, box 361, folder 19.08r7, Regional Programs Joint Program 1962 thru … 1958 [2 of 2].

53. Guided Reponses to questions about the functioning of Euratom, undated, unsigned, NARA, RG 59, box 365, folder 19.8, Regional Program Euratom, General 1958, part 2 of 3.

54. The WEU's membership comprised the UK and the Six. One of its special agencies controlled the importation and production of armament stocks, including fabrication of finished material and components. As explained in undated, unsigned document, NARA, RG 59, box 365, folder 19.08, Regional Programs General June–Dec 1957 thru …1958 [2 of 3].

55. Testimony of Governor Herter with Respect to the Safeguard Arrangements of the Joint US-Euratom Program, July 1958, NARA, RG 59, box 365 folder 19.8, Regional Program Euratom General 1958, part 1 of 2.

56. Georges-Henri Soutou, "Les accords de 1957 et 1858: vers une communauté stratégique nucléaire entre la France, l'Allemagne et l'Italie?" Materiaux pour une histoire de notre temps, no. 31 (1993), 1–12.

57. Soutou, "Les accords," 8 (my translation).

58. Telegram from Secretary of State to US embassy in Belgium, May 24, 1956, *FRUS*, 1955–57, 4: 442–444.

59. H. L. Nieburg, "Euratom: A Study in Coalition Politics," *World Politics* 15 (1963): 597–622, at 599.

60. Memorandum from Elbrick and Smith to Dulles, Designation of Euratom Wise Men and Visit to the United States, December 3, 1956, *FRUS*, 1955–1957, 4: 491–495.

61. John Krige, "Euratom and the IAEA: The Problem of Self-Inspection," *Cold War History*, 15:3 (2015): 341–352.

62. NARA, RG 59 box 358, folder 19.8p, Safeguards part 1, April 10, 1958, memo, Schaetzel, "Safeguards and the Euratom Joint Program."

63. Shane J. Maddock, *Nuclear Apartheid: The Quest for American Atomic Supremacy from World War II to the Present* (University of North Carolina Press, 2010).

64. Memo, Ekern to Farley, "Euratom Safeguards," April 24, 1958, NARA, RG 59, box 358, folder 19.08n, Regional Programs Relations United Kingdom 1958 thru … Peaceful Uses 1958, [1 of 3].

65. Letter, Cole to Eisenhower, May 15, 1958; letter Cole to Strauss, May 12, 1958, NARA, RG 59 box 365 folder Regional Program Euratom General 1958, part 2 of 3.

66. Telegram from London to the secretary of state, June 12, 1958, NARA, RG 59 box 365, folder 19.8, Regional Program Euratom General 1958, part 1 of 2.

67. Memo, Meyers, US embassy London, to Department of State, "British Negotiations with Euratom," June 23, 1958, NARA, RG 59 box 365, folder 19.8, Regional Program Euratom General 1958, part 1 of 2.

68. Memo, Farley to under secretary, "Questions Admiral Strauss may raise…," June 5, 1957, NARA, RG 59, box 365, folder 19.8, Regional Programs Euratom General 1958, part 1 of 2.

69. Telegram from Barbour (London) to Dulles, June 12, 1958, NARA, RG 59, box 365, folder 19.8, Regional Program Euratom General 1958, part 1 of 2.

70. Schaetzel, Testimony of Mr. Dillon With Respect to the Safeguard Arrangements of the Joint US-Euratom Program, July 9, 1958, NARA, RG 59, box 360, folder 19.08r4, Regional Programs Fact Sheets 1953–1960 thru … Joint Program, 1961. Cole's assurance that no one from the communist world would be an inspector as long as he was DG did not persuade.

71. Document EUR/C/103/58 e/es rebutting objections to Euratom Treaty, NARA, RG 59, box 365 folder 19.08, Regional Programs General June–Dec 1957 thru 1958 [2 of 3].

72. Executive Branch Statements Concerning Role of IAEA in Developing Internationally Accepted Safeguards, undated, unsigned, NARA, RG 59, box 365, folder Regional Program Euratom General 1958, part 3 of 3.

73. Confidential. Points to be Raised, undated, unsigned, but drafted by US IAEA officials, NARA, RG 59, box 365, folder Regional Program Euratom General 1958, part 3 of 3.

74. Letter, Cole to Strauss, May 12, 1958, NARA, RG 59, box 365, folder 19.8, Regional Program Euratom General 1958, part 2 of 3.

75. Letter, Cole to President, May 15, 1958, NARA, RG 59, box 365, folder 19.8, Regional Program Euratom General 1958, part 2 of 3.

76. Letter, Strauss to Cole, May 23, 1958, COLE, box 1, folder 9.

77. Telegram from IAEA (Cole) to Strauss, May 29, 1958, COLE,box 2, folder 97.

78. *New York Times*, June 8 and June 9, 1958 (the latter titled "Dispute on Atomic Inspection"), NARA, RG 59, box 365, folder 19.8, Regional Programs General Euratom 1958, part 1 of 2.

79. For an Editorial Note describing the series of meetings and presenting the texts of the agreements, see *FRUS*, 1958–1960, 7: 45–47.

80. The Final Draft of the Agreement, May 6, 1958, is in NARA, RG 59, box 365, folder 19.8, Regional Programs Euratom General 1958, part 1 of 2.

81. Telegram from Dulles to US embassy Brussels and Luxembourg, June 12, 1958, NARA, RG59, box 365, Regional Program Euratom General 1958, part 1 of 2. See also AEC document Joint US-Euratom Cooperative Program. Report by the General Manager, undated, unsigned, NARA, RG 59, box 361 folder 19.08r7, Regional Programs Joint Program 1962, thru … 1958 [2 of 2] (emphasis added).

82. Dulles, telegram, June 12, 1958 (emphasis added).

83. Letter, Baruch to Eisenhower, June 9, 1958, Eisenhower to Baruch, June 13, 1958, cited in Grégoire Mallard, Atomic Confederacy: Europe's Quest for Nuclear Weapons and the New World Order, PhD thesis, Princeton University, 2008.

84. Telegram from Caccia, UK embassy, Washington, to Foreign Office, London, June 13, 1958, UKNA, File AB16/3710.

85. Richard G. Hewlett and Jack M. Holl, *Atoms for Peace and War, 1953–1961: Eisenhower and the Atomic Energy Commission* (University of California Press, 1989), 444.

86. (Draft) letter from Eisenhower to Cole, undated, NARA, RG 59, box 357, folder 19.08j, Regional Programs Documents 1958 thru … OEEC, 1960.

87. Letter, Kohnstamm to Strauss, May 7, 1958, NARA, RG 59 box 357, folder 19.08j, Regional Programs Documents 1958 thru … OEEC 1960.

88. Schaetzel to Butterworth, August 14, 1958, NARA, RG 59, box 361, folder 19.8r, US-Euratom Program Legislation 1958, part 2 of 2.

89. W. Walton Butterworth, Ambassador to Department and AEC, February 7, 1961, NARA, RG 59, box 361, folder 19.8r, US-Euratom Program 8. Legislation 1959–1961.

90. See also memo from Farley to Dillon, Euratom Program Difficulties, February 15, 1960, *FRUS*, 1958–1960, VII: 248–249.

91. Letter, Schaetzel to Butterworth, July 28, 1958 NARA, RG59, box 361, folder 19.8r, US Euratom Program 8. Legislation 1958, part 2 of 2.

92. Winand, *Eisenhower, Kennedy, and the United States of Europe*, 108.

93. Trachtenberg, *A Constructed Peace*, 233.

94. Skogmar, *The United States and the Nuclear Dimension of European Integration*, 256–257.

95. See, for example, Itty Abraham, "The Ambivalence of Nuclear Histories," in *Global Power Knowledge. Science and Technology in International Affairs*, ed. John Krige and Kai-Henrik Barth (*Osiris* 21, 2006); Stephen M. Meyer, *The Dynamics of Nuclear Proliferation* (University of Chicago Press, 1984); Scott D. Sagan, "Why Do States Build Nuclear Weapons? Three Models in Search of a Bomb," *International Security* 21:3 (1996–97): 54–86.

96. Skogmar, *The United States and the Nuclear Dimension of European Integration*, 3.

97. Grégoire Mallard, "L'Europe Puissance Nucléaire, Cet Obscure Objet du Désir," *Critique Internationale* 42 (2009): 141–163; Grégoire Mallard and Andrew Lakoff, "How Claims to Know the Future are Used to Understand the Present," in *Social Science in the Making*, ed. Charles Camic, Michèle Lamont, and Neil Gross (University of Chicago Press, 2011).

98. Mallard, *Fallout* (lies 35, charade 153, opacity 151).

99. Ibid., 3.

100. Testimony of Governor Herter.

101. Euratom Briefing Questions, undated, unsigned, NARA, RG59, box 365, folder 19.8, Regional Program Euratom General 1958, part 2 of 3.

102. Letter, Schaetzel to Butterworth, August 24, 1958, NARA, RG59, box 361, folder 19.8r, US Euratom Program 8. Legislation, 1958, part 2 of 3.

103. John Krige, "Technological Leadership and American Soft Power," in *Soft Power and US Foreign Policy: Theoretical, Historical and Contemporary Perspectives*, ed. Inderjeet Parmar and Michael Cox (Routledge, 2010).

Chapter 3

1. Niklas Reinke, *The History of German Space Policy: Ideas, Influences and Interdependence, 1923–2002* (Beauchesne, 2007), 106–113 (translated from the German).

2. White House press release, "Exchange of Toasts" between Johnson and Erhard, December 20, 1965, LBJL, National Security File, Country File Europe and USSR, Germany, box 192, folder Germany. Erhard Visit [12/65], 12/19–21/65.

3. Hal Brands, "Non-Proliferation and the Dynamics of the Middle Cold War: The Superpowers, the MLF and the NPT," *Cold War History* 7:3 (2007): 389–423, at 397.

4. White House press release, Remarks of the President at Vehicle Assembly Building, Cape Kennedy, Florida, September 27, 1966, LBJL, National Security File, Country File Europe and USSR, Germany, box 193, folder Germany 9/66, Erhard Visit, Papers, Cables, Memos, [9/66].

5. Thomas Alan Schwartz, *Lyndon Johnson and Europe: In the Shadow of Vietnam* (Harvard University Press, 2003). Schwartz has a picture of a rather glum Erhard at the Cape (131) and writes that "Johnson even took Erhard to Cape Kennedy for a launch of a Gemini manned space mission, although this gesture may have only underlined the economic gap between the two nations" (132). In fact no launch coincided with the visit. Two Gemini astronauts, Conrad and Gordon, met Erhard, but neither was on the two Gemini missions that had flown in December of 1965, just days before the chancellor's visit. The next Gemini launch was scheduled for March of 1966.

6. Francis J. Gavin, *Nuclear Statecraft: History and Strategy in America's Atomic Age* (Cornell University Press, 2012), 76.

7. Email message, Wolfgang Finke to the author, August 1, 2007.

8. William Burr and Jeffrey T. Richelson, "Whether to 'Strangle the Baby in the Cradle': The United States and the Chinese Nuclear Program, 1960–1964," *International Security* 25:3 (2000–2001): 55–99. See also the recently declassified documents available at http://www2.gwu.edu/~nsarchiv/nukevault/ebb488.

9. Quoted in Gavin, *Nuclear Statecraft*, 81.

10. A Report to the President by the Committee on Nuclear Proliferation, January 21, 1965, *FRUS*, 1964–1968, 11: 173–182 (The Gilpatric Report).

11. This paragraph is derived from Gavin, *Nuclear Statecraft*, 81ff. Gavin gives a detailed account of the internal differences over these policy options.

12. Gilpatric Report.

13. Ibid.

14. Non-Proliferation of Nuclear Weapons, for Members of the Committee on Nuclear Weapons, Document 27, October 30, 1964 (http://www2.gwu.edu/~nsarchiv/nukevault/ebb488).

15. This connection is not usually made.

16. Gilpatric Report, 10, 12.

17. Scott D. Sagan, "Why Do States Build Nuclear Weapons? Three Models in Search of a Bomb," *International Security* 21:3 (1996–97): 54–86.

18. Pierre Gallois, *Stratégie de l'âge nucléaire* (Calmann-Lévy, 1960). I thank Mark Kramer for this reference.

19. Pierre Messmer, *Journal Officiel*, January 24, 1963, 1613.

20. Gabrielle Hecht, *The Radiance of France: Nuclear Power and National Identity After World War II* (MIT Press, 2009); Walter A. McDougall, "Space Age Europe: Gaullism, Euro-Gaullism and the European Dilemma," *Technology and Culture* 26:2 (1985): 179–203.

21. Robert McNamara, Memorandum for the Vice-President, Brief Analysis of Department of Defense Space Program Efforts, with attached Resume of Existing Programs, April 21, 1961, LBJL, Vice Presidential Security File, box 17, cited by John M. Logsdon, *John F. Kennedy and the Race to the Moon* (Palgrave Macmillan, 2010), 85.

22. Memorandum for the Record, International Space Cooperation, March 23, 1966, *FRUS*, 1964–68, 34: 86.

23. Footnote to Memorandum on International Space Cooperation, March 23, 1966, citing a letter from Webb to McNamara dated April 23, 1964.

24. Letter, Webb to U. Alexis Johnson, May 19, 1966, NARA, RG59, box 3140, Science, folder Space and Astronautics I.

25. Department of State Memo, Science in India; Indo-US and Indo-Pak Relations, October 17, 1966, NARA, RG59, box 3106, folder Science and Technology India

26. Arms Control and Disarmament Agency, Space Cooperation with Japan: Arms Control Considerations, September 6, 1966. Cited, without source, by John M. Logsdon, Learning from the Leader: The Early Years of US-Japanese Cooperation in Space, undated report, Space Policy Institute, Elliot School of International Affairs, George Washington University.

27. Airgram from US embassy in Tokyo to Department of State, Japanese Request for Assistance in Satellite Project, November 27, 1964, NARA, RG59, box 3141, Science, folder Space and Astronautics J.

28. Yasuhiro Nakasone, "Promotion of Science and Technology for Peace" article in *Mainichi Shimbun*, enclosed with airgram from US embassy in Tokyo, Japan's Space Program—The Views of Nakasone, December 15, 1966, NARA, RG59, box 3141, Science, folder Space and Astronautics J.

29. George C. McGhee, Memorandum for the President, Visit of Chancellor Erhard, December 19–21, 1965, LBJL, National Security File, Country File Europe and USSR, Germany box 192, folder Germany, Erhard Visit [12/65], 12/19–21/65.

30. For summary accounts of the MLF see Brands, "Non-Proliferation and the Dynamics of the Middle Cold War"; Maddock, *Nuclear Apartheid*, chapter 8; Trachtenberg, *A Constructed Peace*.

31. Trachtenberg, *A Constructed Peace*, 314.

32. Brands, "Non-Proliferation and the Dynamics of the Middle Cold War," 393.

33. Gilpatric Report, 176.

34. Schwartz, *Lyndon Johnson and Europe*, 61–63.

35. Memorandum of conversation, Nuclear Sharing, December 20, 1965, *FRUS*, 1964–1968, 13: 289–292.

36. The Nuclear Question, Translation of Erhard's Note, December 20, 1965, LBJL, National Security File, Country File Europe and USSR/Germany, box 192, folder Germany Erhard Visit [12/65], 12/19–21/65.

37. Letter, Johnson to UK Prime Minister Wilson, December 23, 1965, *FRUS*, 1964–1968, 13: 295–296.

38. McGhee, Memorandum for the President, Visit of Chancellor Erhard, December 19–21, 1965.

39. Memorandum of conversation, Space and the Great Society, Johnson and Erhard, December 20, 1965, LBJL, National Security File, Country File Europe and USSR, Germany, box 192, folder Germany. Erhard Visit [12/65], 12/19–21/65.

40. Quoted by Reinke, *The History of German Space Policy*, 122.

41. Science, Government and Space Research, Speech made by Ambassador George McGhee at the University of Bonn on January 14, 1966, attached to memo R. Ashton (the Scientific Attaché in the UK embassy in Bonn) to McAdam Clark, January 19, 1966, UKNA, FO 371/189518.

42. Donald Hornig, Memorandum for the President. Visit to Germany and Opportunities for Cooperation, January 31, 1966, LBJL, Papers of D. Hornig, Director, OST, box 3, folder Hornig Chrono File, Jan-Feb 1966.

43. NSAM 357, The Technological Gap, November 25, 1966, LBJL, NSF Subject File box 46, folder Technological Gap (www.lbjlib.utexas.edu/johnson/archives.hom/NSAMs/nsam357.asp).

44. Memo, Hornig to Johnson, Measures to Attack the "Technological Gap" between the United States and Europe, June 7, 1966, *FRUS*, 1964–68, 34: 4–5

45. Arnold W. Frutkin, "The United States Space Program and its International Significance," *Annals of the American Academy of Political and Social Science* 366 (July 1966): 89–98, at 90. See also *Space Business Daily* 25:35 (1966): 285–286.

46. For more details on this see Krige et.al, *NASA in the World*, chapter 2. For Frutkin's frustrations with the British see Letters Smart (British embassy, Washington) to Collins and to Willan (Scientific Relations, the Foreign Office), July 1 and August 10, 1966, UKNA, FO 371/189519.

47. T. W. Garvey, cover note to a report on a meeting at the Foreign Office with James Webb, September 13, 1966, UKNA, FO 371/189519.

48. This was confirmed in the joint communiqué issued by White House after Erhard's visit, September 27, 1966, LBJL, NSF Country File Europe and USSR Germany box 193, folder Germany 9/66, Erhard Visit, Papers Cables, Memos [9/66].

49. Schwartz, *Lyndon Johnson and Europe*, 116.

50. The best account of this complex issue is in chapter 8 of Hubert Zimmerman, *Money and Security: Troops Monetary Policy, and West Germany's Relations with the United States and Britain, 1950–1971* (Cambridge University Press). This quotation is from page 168.

51. Memorandum for the President, Your Meeting with Chancellor Erhard, September 26–27, September 22, 1966. LBJL, National Security File, Country File Europe and USSR, Germany, box 193, folder Germany 9/66, Erhard Visit, Papers, Cables, Memos, [9/66].

52. Text of Cable from Ambassador McGhee (Bonn 3361). The Offset and American Troop Level in Germany, September 20, 1966, LBJL, National Security File, Country File Europe and USSR, Germany, box 193, folder Germany 9/66, Erhard Visit, Papers, Cables, Memos, [9/66].

53. Memoranda to the President from Rostow (September 20, 1966) and McNamara (September 19, 1966), LBJL, NSF, Country File box 193, folder Germany 9/66 Erhard Visit—Papers Cables Memos [9/66].

54. Memorandum of conversation, Offset and Troop Levels. Erhard and Johnson. September 26, 1961. *FRUS*, 1964–1968, 13: 471–477.

55. The White House, Remarks by the President at Vehicle Assembly Building, Cape Kennedy, Florida, September 20, 1966, LBJL, NSF, Country File Europe & USSR, Germany 9/66, Erhard Visit, folder Papers, Cables, Memos [9/66].

56. James E. Webb to Dean Rusk, October 14, 1966, NASA Historical Reference Collection, Washington, Record no.14465, International Cooperation and Foreign Countries, Foreign Countries, West Germany, folder Germany (West), 1956–1990.

57. Webb to Rusk, October 14, 1966.

58. Zimmerman, *Money and Security*, 200.

59. Memo, Eugene V. Rostow to the Secretary, February 8, 1967, LBJL, NSF, Country File/Europe and USSR/Germany, box 193,folder Visit of Vice-Chancellor Brandt, 2/67.

60. Reinhard Loosch, interview with John Krige, Bonn, June 29, 2007, NASA Historical Reference Collection, NASA Headquarters, Washington.

61. The American Team's Visit to Discuss US/European Space Collaboration, attached to unsigned Memo, February 4, 1966, UKNA, FO 371/189518.

62. See Helios Solar Probe at http://www.deutsches-museum.de/en/collections/transport/astronautics/satellites/helios-probe.

Chapter 4

1. Aide-mémoire, Reasons for Terminating the [UKs Commitment …] to ELDO, February 12, 1966, UKNA, CAB 164/7.

2. For a nuanced analysis of this issue see N. Piers Ludlow, "From Words to Actions: Reinterpreting de Gaulle's European Policy," in *Globalizing de Gaulle: International Perspectives on French Foreign Policies, 1958–1969*, ed. Christian Nuenlist, Anna Locher, and Garret Martin (Lexington Books, 2010).

3. Letter, de Gaulle to Johnson, March 7, 1966, *FRUS*, 1964–68, 13: 325, reads "France proposes to recover the entire exercise of her sovereignty over her territory, presently impaired by the permanent presence of allied military elements or by constant utilization which is made of her air space, to terminate her participation in 'integrated' commands and no longer to place her forces at the disposal of NATO."

4. Marie-Pierre Rey, "De Gaulle, French Diplomacy and Franco-Soviet Relations as Seen from Moscow," in *Globalizing de Gaulle*, ed. Nuenlist et al., 35.

5. Quoted by Schwartz, *Lyndon Johnson and Europe*, 104.

6. George Ball, telegram from the Department of State to embassies in Germany and in the other member states of ELDO, April 20, 1966, *FRUS*, 1964–68, 34: 90.

7. Sir P. Dean, telegram to the Foreign Office, May 19, 1966, UKNA, FO371/189518.

8. "Memorandum for the Files. Cooperation with ELDO," May 6, 1966, in LBJL, NSF/Charles Johnson File, box 14, folder Cooperation in Space. Working Group on Expanded International Cooperation in Space, ELDO #2 [2 of 2].

9. Matthew Fargo, "Ballistic Missile Technology 101—Rocket Fuel," at http://csis.org/blog/ballistic-missile-technology-101-rocket-fuel.

10. For India and Japan see Krige et al., *NASA in the World,* chapters 9–12.

11. Stephen R. Twigge, *The Early Development of Guided Weapons in the United Kingdom, 1940–1960* (Routledge, 1993), 347.

12. Ibid., 353. For a full account of the birth of ELDO see John Krige and Arturo Russo, A History of the European Space Agency, 1958–1987, volume I: The Story of ESRO and ELDO, 1958–1973, report SP-1235, European Space Agency, 2000, chapter 3.

13. Notes on Meeting of Jenkins with Stoltenberg, January 5, 1966, UKNA, CAB 164/7.

14. Michael Neufeld, *Von Braun: Dreamer of Space, Engineer of War* (Vintage, 2008) is the most comprehensive biography available.

15. For a detailed chronology and analysis see Michael D. Bordo, Ronald MacDonald, and Michael J. Oliver, Sterling in Crisis, 1964–1967, NBER working paper 14657, 2009.

16. David Edgerton, *Warfare State: Britain 1920–1970* (Cambridge University Press, 2006), chapter 6.

17. P. Rogers, Memo to the Prime Minister, "European Launcher Development Organization (ELDO)," January 14, 1965, UKNA, PREM13/1177.

18. Draft memo addressed to the Foreign Secretary, ELDO, May 27, 1965, UKNA, EW25/52.

19. Cabinet GEN, 716/2nd Meeting, Blue Streak, Minutes of a Meeting … on July 13, 1960, UKNA, CAB130/173, dated July 14, 1960.

20. Department of Defense. International Cooperation in Space Activities, December 16, 1966, Annex A to Summary Meeting of 9th Meeting of the Working Group on Expanded International Cooperation in Space Activities, December 16, 1966 dated February 28, 1967, LBJL, NSF/Charles Johnson File, box 14, folder Cooperation in Space: Working Group on Expanded International Cooperation in Space, ELDO #2.

21. Draft memo addressed to the Foreign Secretary, ELDO, May 27, 1965, UKNA, EW25/52. See also secret memo, Patrick Jefferies to Mr. Neild and others, May 20, 1965, UKNA, EW25/52.

22. P. Rogers, Memo to the Prime Minister, "ELDO," (MISC. 31/2nd Meeting), March 26, 1965, UKNA, PREM13/1177 (underscore for emphasis in original). See also Draft Memorandum by the First Secretary and the Chancellor of the Exchequer, "ELDO," March 20, 1965, UKNA, EW25/52.

23. Conclusions of a Meeting of the Cabinet ... on 16 December 1965, UKNA, CAB/128/39 (www.nationalarchives.gov.uk/cabinetpapers/).

24. Minutes of the 4th Meeting of the Ministerial Conference held on April 28, 1966, HAEU, ELDO/CM(Apr.66)PV/4, April 28, 1966.

25. Europe, Space Policy and ELDO. Report by Officials, January 11, 1966, UKNA, FO378/189508.

26. Letter, T. W. Garvey to Philip Rogers, January 26, 1966, UKNA, CAB164/7. The language referred to Italy in particular, but was generally applicable.

27. T. W. Garvey, ELDO. Brief for Secretary of State, February 2, 1966, UKNA, FO371/189509.

28. Plowden Report on the British Aircraft Industry, February 1, 1966, UKNA, CAB 164/7. For a more general discussion, see David Edgerton, "The 'White Heat' Revisited: The British Government and Technology in the 1960s," *Twentieth Century British History* 7:1 (1996): 53–82.

29. ELDO and Europe (Memorandum by Minister of Aviation), MISC 110(66)11, April 20, 1966, UKNA, FO 371/189512.

30. Aide-mémoire, Reasons for Terminating the [UKs Commitment ...] to ELDO, February 12, 1966, UKNA, CAB 164/7.

31. ELDO and Europe, April 20, 1966.

32. Brief for the Ministerial Conference on ELDO. Minutes of an ad hoc Meeting of Ministers ... on 18 April, 1966; letter from Zuckerman to Minister of Aviation, April 19, 1966; T. W. Garvey, ELDO and "Military" Satellites, April 20, 1966' all in UKNA, FO371/189512.

33. Telegram from Foreign Office (Garvey) to British ambassador in Paris, April 25, 1966, UKNA, FO371/189512.

34. A report on the April Ministerial meeting is attached to C(66)73, Memorandum by the Minister of Aviation on ELDO for the Cabinet, May 24, 1966, UKNA, CAB129/125 (www.nationalarchives.gov.uk/cabinetpapers/).

35. Outgoing telegram 1209, Department of State to American embassy in Bonn, November 18, 1965, signed [George] Ball, LBJL, National Security File, box 192, folder Germany Erhard Visit [12/65], 12/19–21/65, and folder Germany, Country File Europe and USSR.

36. Notes on Meeting between Pollock and Zuckerman, Washington, 6 May, attached to memo for the Minister of Aviation, May 10, 1966.

37. This issue so exercised the Johnson administration that Johnson himself established a special task force to look into it. The directive (The Technological Gap, NSA357, November 25, 1966, signed by the president himself) is available at http://www.lbjlib.utexas.edu/johnson/archives.hom/nsams/nsam357.asp. The lengthy interim report that emerged a year later was titled The Nature and Causes

of the Technological Gap Between Western Europe and the United States. November 1967, in LBJL, National Security File, Subject File, box 46, folder Technological Gap [2 of 2].

38. Meeting Between Mr. Webb, NASA, and Sir Solly Zuckerman, Washington, 6 May, attached to memo for the Minister of Aviation, May 10, 1966, UKNA, CAB164/9.

39. Memo, T. W Garvey to William S. Gruelich, US embassy, May 19, 1966, UKNA, FO371/189518.

40. Memo, T. W. Garvey for J. E. Killick, Washington, ELDO: US Interest, May 6, 1966, reports on the meeting with Frutkin in London, UKNA, FO371/189518.

41. Note for the Record on Zuckerman's report to the Prime Minister on May 9, 1966, undated, unsigned, UKNA, PREM 13/1178.

42. Memo, Burke Trend to the Prime Minister on ELDO, May 25, 1966, UKNA, PREM13/1178.

43. NSAM294 is available at www.lbjlib.utexas.edu/johnson/archives.hom/nsams/nsam294.asp.

44. A Report to the President by the Committee on Nuclear Proliferation, January 21, 1965, *FRUS*, 1964–1968, 11: 173–182 (The Gilpatric Report), emphasis added.

45. NSAM338, Policy Concerning US Assistance in the Development of Foreign Communications Satellite Capabilities, September 15, 1965 (http://fas.org/irp/offdocs/nsam-lbj/nsam-338.htm).

46. Policy Concerning US Assistance in the Development of Foreign Communications Satellite Capabilities, August 25, 1965, attached to memo J.D. O'Connell to Secretary of State et al., September 17, 1965, LBJL, NSF/NSAM Files, box 7, folder NSAM 338. This folder contains extensive material on US policy in this area. See also Krige et al., *NASA in the World,* chapter 5.

47. Policy Concerning US Assistance, 16.

48. McGeorge Bundy, Memorandum for the President, Helping Others to Use Communications Satellites, September 13, 1965, LBJL, NSF/NSAM Files, box 7, folder NSAM 338 (underscore for emphasis in original).

49. Memo by R. F. Packard, Conditions Underlying Cooperation in the Development of Foreign Communications Satellite Capabilities. Discussion Paper, August 4, 1966, LBJL, NSF/Charles Johnson File, box 14, folder Cooperation in Space, Working Group on Expanded International Cooperation in Space, ELDO #1 (1 of 2).

50. Note for the Record from Zuckerman to the Prime Minister, May 9, 1966, UKNA, PREM 13/1178.

51. T. W. Garvey, ELDO and "Military" Satellites, April 20, 1966, UKNA, FO371/189512. In an agreement signed in September of 1966, the US undertook to construct and launch two spin-stabilized communications satellites paid for by the UK and for its sole use. But they would be part of a constellation comprising the Department of Defense's communications satellite program, and would be

used by Britain "for those communications intimately concerned with the defense of the Realm."

52. Memo by the Minister of Aviation on ELDO and Europe in anticipation of the ELDO Ministerial Meeting, May 24, 1966, UKNA, CAB129/125 (www.nationalarchives.gov.uk/cabinetpapers).

53. Conclusions of a Meeting of the Cabinet ... on 26th May, 1966, UKNA, CAB128/41 (www.nationalarchives.gov.uk/cabinetpapers).

54. Ibid.

55. Memos C. O'Neill (31 May, 1966), P. H. Gore-Booth (June 1, 1966) and Lord Chalfont (June 1, 1966), UKNA, FO371/ 189515.

56. Telegram from Sir R. Barclay from Brussels to the Foreign Office, June 6, 1966, UKNA, PREM13/1178.

57. Letter to Prime Minister from Society of British Aerospace Companies, Ltd., June 7, 1966, UKNA, CAB 164/44.

58. C(66)75, Memorandum by the Minister of Aviation to the Cabinet, June 8, 1966, and Annex A, UKNA, CAB 129/125.

59. Conclusions of a Meeting of the Cabinet [held on] 9th June, 1966, UKNA, CAB 128/41.

60. Ministerial Conference in Paris on 9/10th June, 1966, Note by the Minister of Aviation, June 23, 1966, UKNA, CAB 129/125. See also Memorandum by the Minister of Aviation for the Cabinet, July 20, 1966, UKNA, CAB 129/126, (www.nationalarchives.gov.uk/cabinetpapers).

61. This was particularly onerous for the FRG, whose share of the ELDO budget increased from 22 percent to 27 percent.

62. Memorandum of the Minister of Aviation for the Cabinet, July 20, 1966, UKNA, CAB129/126 (www.nationalarchives.gov.uk/cabinetpapers).

63. NSAM 354, US Cooperation with the European Launcher Development Organization, July 29, 1966 (www.lbjlib.utexas.edu/johnson/archives.hom/NSAMs/nsam354.asp).

64. US Cooperation with ELDO, July 21, 1966, LBJL, NSF/Charles Johnson File, box 14, folder Cooperation in Space. Working Group on Expanded International Cooperation in Space. ELDO #1 [2 of 2].

65. Report to the National Aeronautics and Space Council concerning Opportunities for Expanded International Cooperation in Outer Space Activities, February 28, 1967, 27, LBJL, NSF/Charles Johnson Files, box 14, folder Cooperation in Space: Working Group on Expanded International Cooperation in Space (ELDO) #2.

66. Policy Concerning US Cooperation with the European Launcher Development Organization (ELDO), attached to NSAM354, July 20, 1966, LBJL, NSF/Charles Johnson Files, box 14, folder Cooperation in Space: Working Group on Expanded International Cooperation in Space (ELDO) #1, [1 of 2].

67. Letter, Richard Barnes to Scott George, April 15, 1966, LBJL, NSF/Charles Johnson Files, box 14, folder Cooperation in Space, Working Group on Expanded International Cooperation in Space, ELDO #2 [2 of 2].

68. They were not to advance nuclear missile delivery capabilities of any of the member states either, of course. R. F. Packard, Discussion Paper, Conditions Underlying Cooperation in the Development of Foreign Communications Satellites Capabilities, August 4, 1966, LBJL, NSF/Charles Johnson Files, box 14, folder Cooperation in Space: Working Group on Expanded International Cooperation in Space. ELDO. #1 [1 of 2].

69. Policy Concerning US Cooperation with the European Launcher Development Organization (ELDO), attached to NSAM354.

70. Policy Concerning US Cooperation with the European Launcher Development Organization (ELDO), attached to NSAM354.

71. Packard, Discussion Paper, August 4, 1966.

72. Communications Satellite Technology, undated, unsigned, LBJL, NSF/Charles Johnson Files, box 14, folder Cooperation in Space: Working Group on Expanded International Cooperation in Space (ELDO) #2 [2 of 2].

73. Letter, U. Alexis Johnson to James O'Connell, September 3, 1966, LBJL, NSF/Charles Johnson Files, box 14, folder Cooperation in Space: Working Group on Expanded International Cooperation in Space (ELDO) #2.

74. ELDO/CM(July 68) WP/2, Cooperation with the United States, July 10, 1967, HAEU, Florence, ELDO, July 1967, ELDO-977, Cooperation with USA.

75. As clearly explained by Trevanion H. E. Nesbitt in Germany on June 21, 1968. See Annex to ELDO/CM(July 68)WP/2.

76. Summary minutes of meeting between Ortoli, Leussink (FRG) and Lefèvre (B), on June 1, 1971, reproduced in *L'ambition technologique: naissance d'Ariane*, ed. Emmanuel Chadeau (Rive Droite, 1995), 406 (my translation).

77. Opportunities for Expanded International Cooperation in Outer Space Activities. A Report to the National Aeronautics and Space Council from its Inter-agency Working Group on Expanded International Cooperation, May, 1967, LBJL, NSF/Charles Johnson Files, box 14, folder Cooperation in Space. Working Group on Expanded International Cooperation in Space, ELDO #1 [1 of 2], 5–6.

78. Ibid., iv.

79. Trevanion H. E. Nesbitt, "Possibilities and Problems of Future US-European Cooperation in the Space Field," Remarks at a Eurospace meeting in Munich, Germany, June 21, 1968, annex to ELDO/CM(July 68) WP/2, Cooperation with the United States, July 10, 1967, HAEU, Florence, ELDO, July 1967, ELDO-977, Cooperation with USA.

80. ELDO/CM(Apr.66)PV/2, Annex, April 27, 1966, HAEU.

81. N. Piers Ludlow, "European Integration and the Cold War," in *The Cambridge History of the Cold War*, volume 2: *Crises and Détente*, ed. Melvyn P. Leffler and Odd Arne Westad (Cambridge University Press, 2010), 193.

Chapter 5

1. Luc Gillon, "L'approvisionnement en uranium," in *L'Énergie nucléaire en Europe: Des origins à Euratom*, ed. Michel Dumoulin, Pierre Guillen, and Maurice Vaïsse (Peter Lang, 1994).

2. R. Scott Kemp, "The End of Manhattan. How the Gas Centrifuge Changed the Quest for Nuclear Weapons," *Technology and Culture* 53:2 (2012): 272–305, at 272.

3. Statement by the President Announcing the Use of the A-bomb at Hiroshima (www.trumanlibrary.org/publicpapers/index.php?pid=100&st=atomic&st1=bomb). Also see Kemp, "The End of Manhattan."

4. Gregg Herken, "'A Most Deadly Illusion'. The Atomic Secret and American Nuclear Weapons Policy, 1945–1950," *Pacific Historical Review* 49:1 (1980): 51–76, at 57.

5. This chapter has benefited enormously from discussions with William Burr and from the recently declassified documents he posted: National Security Archive Electronic Briefing Book No. 385, Early Atomic Energy Commission Studies Show Concern over Gas Centrifuge Proliferation Risk, July 23, 2012 (www2.gwu.edu/~nsarchiv/nukevault/ebb385). For other studies of this topic, see John Krige, "The Proliferation Risks of Gas Centrifuge Enrichment at the Dawn of the NPT: Shedding Light on the Negotiating History," *Nonproliferation Review* 19:2 (2012): 219–227; John Krige, "Hybrid Knowledge. The Transnational Coproduction of the Gas Centrifuge for Isotope Enrichment in the 1960s," *British Journal for the History of Science* 45:3 (2012): 337–357; John Krige, "US Technological Superiority and the Special Nuclear Relationship: Contrasting British and US Policies for Controlling the Proliferation of Gas-Centrifuge Enrichment," *International History Review* 36:2 (2014): 230–251; Susanna Schrafstetter and Stephen Twigge, "Spinning into Europe: Britain, West Germany and the Netherlands—Uranium Enrichment and the Development of the Gas Centrifuge, 1964–1970," *Contemporary European History* 11 (2002): 253–272; Stephen Twigge, "A Baffling Experience: Technology Transfer, Anglo-American Nuclear Relations, and the Development of the Gas Centrifuge, 1964–1970," *History and Technology* 19 (2003): 151–163; R. Scott Kemp, Nonproliferation Strategy in the Centrifuge Age, PhD dissertation, Princeton University, 2010; Kemp, "The End of Manhattan"; Houston G. Wood, Alexander Glaser, and R. Scott Kemp, "The Gas Centrifuge and Nuclear Weapons Proliferation," *Physics Today* 61:9 (2008): 40–45.

6. The description of Zippe's trajectory and the subsequent work done at Oak Ridge up to July 1965 is from Kemp, "The End of Manhattan" and R. Scott Kemp, "Gas Centrifuge Theory and Development: A Review of US Programs," *Science and Global Studies* 17 (2009): 1–19.

7. Kemp describes this early period well in "The End of Manhattan."

8. Anthony Wedgwood Benn, *Office without Power: Diaries 1968–72* (Hutchinson, 1988), 127.

9. Gas Centrifuge Method of Isotope Separation, Report AEC610/1, April 9, 1960, NSAE385, Document 6, 2. See also Letter Sullivan to Dowling, April 28, 1960, NSAE518 Document 11.

10. Kemp, "The End of Manhattan," 286.

11. Report AEC 610/15, April 9, 1960, NSAE385, Document 6, 16.

12. Ibid., 30.

13. Union Carbide Nuclear Company, Production of Enriched Uranium for Nuclear Weapons by Nations X, Y and Z by Means of the Gas Centrifuge Process, February 26, 1960, NSAE385, Document 4, 4.

14. Report AEC 610/15, April 9, 1960, NSAE385, Document 6, 4.

15. Letter, Marshall to Wells, December 7, 1959, appendix G to AEC610/15, April 9, 1960.

16. Report AEC 610/15, April 9, 1960, NSAE385, Document 6, 3.

17. JCAE, Meeting 86–2-49, August 30, 1960, To Receive Briefing on Centrifuge Process, NSAE 518, Document 23, 42, 52.

18. Testimony of John A. McCone, JCAE, Meeting 86–2-49, August 30, 1960, Attached.

19. William Burr, "The 'Labors of Atlas, Sisyphus, or Hercules'? US Gas-Centrifuge Policy and Diplomacy, 1954–1960," *International History Review*, 37:3 (2015) 431-457. William Burr, "To 'Keep the Genie Bottled Up': U.S. Diplomacy and Gas Centrifuge Technology, 1962–1972," *Journal of Cold War Studies*, in press.

20. Memorandum of conversation, Herter (State), McCone (AEC) and Farley (AEC), June 8, 1960, NSAE518, Document 14.

21. Burr, "The 'Labors of Atlas, Sisyphus, or Hercules,'" 13.

22. Letter, Farley to Wells, March 23, 1960, NSAE518, Document 11.

23. Telegram to Bonn, London, and The Hague, Gas Centrifuge Classification Guide, NSAE518, Documents 20A and 20B, July 27 and October 19, 1960.

24. Burr, "To 'Keep the Genie Bottled Up.'"

25. Memorandum of conversation, German Classification of Centrifuge Work, October 4, 1960, and Memorandum for the Files, Press Conference in Bonn on Ultracentrifuge Process, October 13, 1969, NSAE518, Document s 25 and 26.

26. PNO (C) (69) 22 (Final), June 2, 1969, Cabinet Official Committee on Nuclear Policy, "Centrifuge Collaboration. Anglo/United States Relations in the Nuclear Field," UKNA, FCO55/268.

27. Summary Report of Nth Power Probability of Producing Enriched Uranium via the Gas Centrifuge Method, attached to letter Seaborg to McGeorge Bundy, November 22, 1965, NSAE385, Document 9.

28. US AEC press release, "AEC Sets New Policy on Gas Centrifuge Development," 21 March 1967, LBJL, NSF, box 33, folder Nuclear.

29. Seaborg to Rostow, Mar 10, 1967, LBJL, NSF, box 33, folder Nuclear.

30. McNamara to Seaborg, May 23, 1964, Nuclear Testing Archive, Las Vegas, NV, document NV0903211, cited http://nuclearsecrecy.com/blog/2012/5/30/weekly-documents-the-centrifuge-conundrum-19641968.

31. Kemp, "Gas Centrifuge Theory and Development," table 1.

32. Letter, Seaborg to McGeorge Bundy, November 22, 1965, NSAE385, Document 9.

33. CH(67)5, The Production of Enriched Uranium, July 26, 1967, UKNA, PREM13/2004, 6.

34. Memo, Benn to Prime Minister, "Acarus" [code name for the centrifuge project], December 20, 1967, UKNA, PREM13/2004.

35. Helen Parr, *Britain's Policy Toward the European Community: Harold Wilson and Britain's World Role, 1964–1967* (Routledge, 2006), 170.

36. Ibid.

37. Ibid., 178.

38. Memo, Sir Solly Zuckerman, February 29, 1968, UKNA, CAB164/732.

39. Burr, "To 'Keep the Genie Bottled Up.'"

40. Memo, Benn to Prime Minister, April 9, 1968, UKNA, CAB164/732.

41. Memo, Benn to Prime Minister, March 4, 1968, UKNA, PREM13/2004.

42. It was to be a "planetary system" comprising thermonuclear weapons carried by bombers (France's first hydrogen bomb was tested in August of 1968), intercontinental ballistic missiles, and submarine launched missiles that could strike "wherever necessary on the face of the earth." Frédéric Bozo, *Two Strategies for Europe: De Gaulle, the United State and the Atlantic Alliance* (Rowman and Littlefield, 2000), 203–204.

43. Memo, Benn to Wilson, European Nuclear Policy, March 14, 1968, UKNA, PREM13/2004.

44. Memo, Benn to Prime Minister, March 4, 1968.

45. Memo, Benn to Wilson, European Nuclear Policy, March 14, 1968.

46. Memo, Denis Healey to Prime Minister, March, 11, 1968, UKNA, PREM13/2004.

47. Solly Zuckerman, Chair of the Working Party, Report on Cooperation with Europe on Nuclear Energy, March 14, 1968, UKNA, FCO55/111.

48. Memo, Zuckerman to the Prime Minister, Cooperation with Europe on Nuclear Energy, March 14, 1968, UKNA, PREM 13/2004.

49. Flash telegram from Foreign Office to Washington, "Uranium Enrichment by the Gas Centrifuge Process," April 10, 1968, UKNA, FCO55/111.

50. Memo 'Non-Proliferation' attached to a memo signed by T. W. Garvey, April 11, 1968, UKNA, FCO55/111.

51. Memo under cover of a memo from T. Garvey, April 11, 1968.

52. Flash telegram from Foreign Office to Washington, April 10, 1968.

53. Copy of telegram from Department of State to Foreign Office, April 15, 1968, NARA, RG59, Entry A1(5618) Lot74D11, box No. 2, folder "Centrifuges and Space Technology Transfer." I assume this is a copy, as it is not annotated. British documents refer to the telegram, but I have not found the original.

54. Unsigned, untitled memo, dated April 15, 1968NARA, folder S.23 Gas Centrifuge, box 22.

55. Gore-Booth to Dean, Washington, "Implications of Gas Centrifuge Development for the Non-Proliferation Treaty," and attached "Talking Points" prepared by the Atomic Energy and Disarmament Department in the Foreign Office, April 29, 1968, UKNA, FCO10/207.

56. Hope-Jones to Peck, "The Centrifuge Process and the UK Safeguards Offer," April 30, 1968, UKNA, FCO10/207.

57. Ibid.

58. Audland to Peck, April 18, 1968, UKNA, FCO10/207; Draft telegram to Washington, April 18, 1968, UKNA, FCO10/207; Letter Falkley to Hope-Jones, April 25, 1968, UKNA, FCO10/207; "Redraft of Paragraphs 7 to 10 of MISC 191(68)4(Revised)," Atomic Energy and Disarmament Department, Foreign Office, undated, but about this time, UKNA, FCO10/207.

59. Audland to Peck, April 18, 1968.

60. Maitland to Hope-Jones, February 27, 1969, UKNA, FCO 55/268.

61. Matthias Küntzel, *Bonn and the Bomb: German Politics and the Nuclear Option* (Pluto, 1995), 90. This is just an example of the stream of invective in the FRG that greeted the final draft of the NPT proposed by the US and the USSR, a collaboration between the superpowers made possible by the United States' acceptance of the USSR's demands that the MLF be abandoned. See also Susanna Schrafstetter, "The Long Shadow of the Past: History, Memory and the Debate over West Germany's Nuclear Status, 1954–69," *History and Memory* 16:1 (2004): 118–145. See also Schrafstetter and Twigge, *Avoiding Armageddon*, 182 ff.

62. Glenn T. Seaborg, *Stemming the Tide: Arms Control in the Johnson Years* (Lexington Books, 1987), 359.

63. *New York Times*, July 18, 1968.

64. Ronald J. Granieri, "Odd Man Out? The CDU-CSU, Ostpolitik, and the Atlantic Alliance," in *The Strained Alliance: US-European Relations from Nixon to Carter*, ed. Matthias Schulz and Thomas A. Schwartz (German Historical Institute and Cambridge University Press, 2010), 93.

65. R. C. Hope-Jones to Audland, "Gas Centrifuge Collaboration," November 29, 1968, UKNA, FCO66/78.

66. M. J. Newington to Killich, "Centrifuge Collaboration: Germany and the N.P.T.," May 21, 1969, UKNA, FCO66/79.

67. Hope-Jones to Audland, November 29, 1968.

68. For this paragraph, see Burr, "To 'Keep the Genie Bottled Up.'"

69. Unsigned, untitled Memo, dated April 15, 1968, NARA, folder S.23 Gas Centrifuge, box 22

70. Telegram from Stewart, Foreign and Commonwealth Office, to Washington, March 14,1969, UKNA, FCO55/265.

71. Summarized, e.g., in Anglo/US Relations in the Nuclear Field, paper prepared for Cabinet Ministerial Committee on Nuclear Policy, Centrifuge Collaboration, May 19, 1969, UKNA, CAB134/314.

72. Telegrams 3552 and 3553 from Washington to Foreign and Commonwealth Office, December 6. 1968, UKNA, PREM13/2555.

73. Telegram 3553, Washington to Foreign and Commonwealth Office.

74. Ibid.

75. For details, see Twigge, "Baffling"; Krige, "Maintaining."

76. Myron Kratzer was the AEC's Assistant General Manager for International Activities. Paul Vanstrum was the Technical Director of Production for Union Carbide Nuclear at Oak Ridge. He had extensive experience of the US gas centrifuge—see Kemp, "Nonproliferation Strategy in the Centrifuge Age," 22.

77. Record of United States/United Kingdom Talks Held at the Cabinet Office, Whitehall, 4, 5 March, 1969, UKNA, FCO55/265; Record of a Meeting on 3 May at the Office of the US Atomic Energy Commission, Washington D.C., UKNA, PREM13/2556.

78. Draft Memorandum. Interpretation of Article IX(c) of the 1955 Civil Bilateral Agreement, PNC(C)(68)12, December 19, 1968, UKNA, CAB 134/3125.

79. Report Anglo/US Relations in the Nuclear Field, PN (69)8, May 19, 1969, Cabinet Ministerial Committee on Nuclear Policy. Centrifuge Collaboration, UKNA, CAB134/3121.

80. "Centrifuge Technology," Record of US/UK talks held on 5 March, 1959, UKNA, FCO 55/265.

81. Zuckerman to Prime Minister, Centrifuge Collaboration. Enquiry by Lord Penney, Sir Alfred Pugsley and Mr. T. C. Hetherington, June 2, 1969, UKNA, PREM13/2556.

82. Memo, Solly Zuckerman to Prime Minister, Centrifuge Collaboration, May 21, 1969, UKNA, PREM 13/2556.

83. Cited by Burr, "To 'Keep the Genie Bottled Up.'"

84. Record of a Meeting on May 3 in Washington.

85. Telegram 1367, Washington to Foreign and Commonwealth Office, May 4, 1969, UKNA, PREM13/2556.

86. PNO(69) 4th Meeting, Cabinet Official Committee on Nuclear Policy, Minutes of a Meeting held in the Cabinet Office on 20 May, 1969, UKNA, FCO 66/79.

87. Memo, Zuckerman to Prime Minister, Centrifuge Collaboration, May 21, 1969, UKNA, PREM 13/2556.

88. Memo, Hill, Centrifuge, May 21, 1969, attached to memo, Dunnett to Zuckerman, Centrifuge, May 21, 1969, UKNA, PREM 13/2556.

89. Report of Enquiry Relating to Restricted Data on Centrifuge Design and Construction … , May 30, 1969, UKNA, FCO55/268.

90. Letter, Zuckerman to Prime Minister, June 2, 1969.

91. British aide-mémoire, July 3, 1969, UKNA, CAB 134/3121.

92. Centrifuge Technology, Record of US/UK talks held on 4 March, 1969.

93. Record of a Meeting on June 6, 1969, at US AEC Office, UKNA, PREM 13/2556.

94. British aide-mémoire, July 3, 1969.

95. The fourth member seems to have been either Ed Babely or Ernest Evans from Oak Ridge or Ralph Lowry from the centrifuge team at the University of Virginia. NARA, RG 39, Entry A1(5618) Lot 74D11, box 2, folder S.23, Gas Centrifuge Technology 1969.

96. Memo, Zuckerman to Prime Minister, Centrifuge. *The Problem of Article IXC of the US/UK Civil Agreement*, July 18, 1969, UKNA, PREM 13/2556.

97. Memo, Zuckerman to Prime Minister, March 4, 1970, UKNA, PREM 13/3128.

98. US Department of State, aide-mémoire, October 1, 1969, UKNA, FCO 55/271.

99. The official history is R. B. Kehoe, *The Enriching Troika: A History of Urenco to the Year 2000* (Urenco, 2000).

100. As Nelson Sievering put it on returning home from his visit to the UK, "if the Oak Ridge analysis substantiates the poor potential of the UK production model, especially the limitations which the current end cap imposes … the sooner we can cut the umbilical cord and make a clean break between the 1960—65 exchanges and the current UK effort the better off we will be." Memo, Nelson Sievering to US AEC Commissioner Theos Thompson, Reflections on the Gas Centrifuge Meetings with the UK, July 31, 1969, NARA, folder S.23 Gas Centrifuge, box 22.

101. Letter, Thompson to Holifield, August 8, 1969, Glenn Seaborg's Office Diary, folder-Page 104139, kindly made available to me by William Burr.

102. Anglo-US Relations in the Nuclear Field, attached to circular by the Secretariat for the Cabinet Official Committee on Nuclear Policy, June 2, 1969, UKNA, FCO55/268.

103. Solly Zuckerman, *Monkeys, Men and Missiles: An Autobiography, 1964–1988* (Collins, 1988), 445.

104. Benn, *Office Without Power*, Documents 4 and 46. 127.

Conclusion

1. For typical arguments, see *Atoms for Peace*, ed. Pilat et al.; Sokolski, *Best of Intentions*.

2. John Baylis, "Exchanging Nuclear Secrets: Laying the Foundations of the Anglo-American Nuclear Relationship," *Diplomatic History* 25:1 (2001): 33–61.

3. In a letter to McNamara dated April 17, 1964, cited in letter McNamara to Seaborg, May 23, 1964, available on Alex Wellerstein's Nuclear Secrecy Blog in the posting "The Centrifuge Conundrum."

4. I thank Alex Wellerstein for drawing my attention to this point.

5. Cited by Geir Lundestad on p. 22 of *"Empire" by Integration.*

6. Meeting of JCAE, August 30, 1960, NSA 518, 22; memo, Benn to Prime Minister, March 4, 1968, UKNA, PREM13/2004.

7. NSAM294, April 20, 1964, US Nuclear and Strategic Delivery System Assistance to France (www.lbjlib.utexas.edu/johnson/archives.hom/NSAMs/nsam294.asp).

8. Memorandum of conversation with Kissinger and others, August 17, 1973, WCDA, Document 43.

9. Memorandum of conversation with Robert Galley, French Minister of the Armed Forces, and Henry Kissinger, July 27, 1973, WCDA, Document 41.

10. Memoranda of conversations between Kissinger and Galley, San Clemente White House, July 27 and August 31, 1973, WCDA Docs.#4, #46.

11. Memorandum of conversation, Dulles with Lloyd, March 23, 1957, Bermuda, WCDA, Document 1.

12. Background study prepared by Packard for Kissinger, US/French Interchange in Area of Ballistic Missiles, February 20, 1970, WCDA, Document 12.

13. Dulles to Lloyd in Bermuda, March 1957. Kennedy persisted with this policy, offering de Gaulle the Polaris weapons system in December of 1962—see John Newhouse, *De Gaulle and the Anglo-Saxons* (Viking, 1970), 224.

14. Richard H. Ullman, "The Covert French Connection," *Foreign Policy* 75 (summer 1989): 3–33, at 17–18.

15. It is something of an irony that the sources that I use in what follows don't deal with debates over how to stop a country from getting a nuclear weapon but rather with how to help it improve the weapons it has. This sits uneasily in a book that is concerned with curbing proliferation, though for the actors involved the purpose of the exchange was to stabilize the world order under American leadership. For a leading power, if you could not retard or stop your allies' getting a bomb, there was all the more need for peering into their program, and keeping an eye on how it evolved so better to control it.

16. Jeremi Suri, "Henry Kissinger and the Geopolitics of Globalization," *The Shock of the Global: The 1970s in Perspective*, ed. Niall Ferguson, Charles Maier, Erez Manela, and Daniel Sargent (Belknap, 2011).

17. Memorandum of conversation, Kissinger with Schlesinger, September 5, 1973, WCDA, Document 47; Helmut Sonnenfeldt, Memorandum for Mr. Kissinger, July 26, 1973, WCDA Document 40.

18. Memorandum of conversation, Kissinger, Schlesinger, and Scowcroft, August 9, 1973, WCDA, Document 42.

19. Memorandum of conversation, Kissinger and others, August 17, 1973, WCDA, Document 43.

20. Memorandum of conversation, Kissinger, Schlesinger, and Scowcroft, August 9, 1973, WCDA, Document 42.

21. Memorandum of conversation, Kissinger and others, August 17, 1973, WCDA, Document 43.

22. Quoted by Marc Trachtenberg, "The French Factor in US Foreign Policy During the Nixon-Pompidou Period, 1969–1974," *Journal of Cold War Studies* 13:1 (2011): 4–59, at 39.

23. Pierre Melandri, "Aux origins de la cooperation nucléaire franco-américaine," in *La France et l'atome: Études d'histoire nucléaire*, ed. Maurice Vaïsse (Bruylant, 1994); Georges-Henri Soutou, "Le President Pompidou et les relations entre les États-Unis et l'Europe," *Journal of European Integration History* 6:2 (2000): 111–146; Maurice Vaïsse, "Les 'relations spéciales franco-américains aux temps de Richard Nixon et Georges Pompidou," *Relations Internationales* 119 (2004): 345–362.

24. On Kissinger's backchannel diplomacy see Tony Judt, "The Illusionist: Henry Kissinger and American Foreign Policy," in *Reappraisals: Reflections on the Forgotten Twentieth Century* (Penguin, 2008), chapter 20.

25. Memorandum of conversation, Galley and Kissinger, August 31, 1973, WCDA, Document 46.

26. Quoted in Maurice Vaïsse, *La Puissance ou l'Influence? La France dans le monde depuis 1958* (Fayard, 2009), 194.

27. Sonnenfeldt, Memorandum for Mr. Kissinger, Assistance to French Ballistic Missile Program, April 16, 1970, WCDA, Document 17.

28. NSSM100—Military Cooperation with France, Report of the National Security Staff, ca. December 1970, WCDA, Document 20B.

29. Kissinger, Memorandum for the President, Military Cooperation with France, March 25, 1971, WCDA, Document 22.

30. NSDM103, Advanced Military Cooperation with France, March 29, 1971 (http://fas.org/irp/offdocs/nsdm-nixon/nsdm-103.pdf).

31. Ibid.

32. Sonnenfeldt, Memorandum for the Record, Conversation with Delpech, October 9, 1975, WCDA, Document 55.

33. Laird to Kissinger, July 29, 1971, WCDA, Document 27.

34. Ibid.

35. NSDM103, Advanced Military Cooperation with France, March 29, 1971.

36. NSDM104, Cooperation with France on Nuclear Safety, March 29, 1971 (http://fas.org/irp/offdocs/nsdm-nixon/nsdm-104.pdf).

37. Ronald I. Spiers to the Deputy Secretary, "Military Cooperation with France: Outcome of the Debre Visit," August 28, 1972, WCDA, Document 35B.

38. Memorandum of conversation, Galley and Kissinger, August 31, 1973.

39. Ullman, "The Covert French Connection," 20.

40. John Baylis, "Exchanging Nuclear Secrets," 59. The person quoted is Frederick Jandrey, Deputy Secretary for European Affairs.

41. Ullman, "The Covert French Connection," 13.

42. As described in Vaïsse, "Les 'relations speciales' franco-américaines'. See also Georges-Henri Soutou, "Georges Pompidou and US-European Relations," in *Between Empire and Alliance: America and Europe During the Cold War*, ed. Marc Trachtenberg (Rowman and Littlefield, 2003).

43. Matthew Fuhrmann, *Atomic Assistance: How "Atoms for Peace" Programs Cause Nuclear Insecurity* (Cornell University Press, 2012). See also Matthew Kroenig, *Exporting the Bomb: Technology Transfer and the Spread of Nuclear Weapons* (Cornell University Press, 2010).

44. Roland Popp, "Introduction: Global Order, Cooperation between the Superpowers, and Alliance Politics in the Making of the Nuclear Non-Proliferation Regime," *International History Review* 36:2 (2104): 195–209.

45. More recently, Siegfried Hecker, a professor at Stanford University and a former director of the Los Alamos National Laboratory, was shown an impressive new centrifuge enrichment plant by the North Korean authorities. See David E. Sanger, "North Koreans Unveil New Plant for Nuclear Use," *New York Times*, November 20, 2010. In a recent email to me, Hecker wrote "You are correct about the importance of technical collaboration—but not just as instruments of surveillance/control, but also of hope. In the case of North Korea, I did not go as an inspector. They had enough of those already. I went to reduce the ambiguities in their nuclear program and to provide a forum in which they could express their hopes."

Bibliography

Abraham, Itty. *The Making of the Indian Atomic Bomb*. London: Zed, 1998.

Abraham, Itty. The Ambivalence of Nuclear Histories. In *Global Power Knowledge: Science and Technology in International Affairs*, ed. John Krige and Kai-Henrik Barth (*Osiris* 21, 2006).

Abraham, Itty. Contra-Proliferation: Interpreting the Meanings of India's Nuclear Tests in 1974 and 1998. In *Inside Nuclear South Asia*, ed. Scott D. Sagan. Stanford University Press, 2009.

Adas, Michael. *Dominance by Design: Technological Imperatives and America's Civilizing Mission*. Cambridge: Belknap, 2006.

Baylis, John. Exchanging Nuclear Secrets: Laying the Foundations of the Anglo-American Nuclear Relationship. *Diplomatic History* 25:1 (2001): 33–61.

Benn, Anthony Wedgwood. *Office Without Power: Diaries 1968–1972*. London: Hutchinson, 1988.

Berghahn, Volker, *America and the Intellectual Cold Wars in Europe*. Princeton University Press, 2002.

Bhabha, Honmi J. The Role of Atomic Power in India and Its Immediate Possibilities. In *Proceedings. Geneva 1955*, volume I: *The World's Requirements for Energy. The Role of Nuclear Energy*. New York: United Nations, 1956.

Blamont, Jacques. Origines et principes de la politique spatiale de la France. *Rayonnement du CNRS* 48:June (2008): 24–32.

Bordo, Michael D., Ronald MacDonald, and Michael J. Oliver. Sterling in Crisis, 1964–1967. NBER working paper 14657 (available at www.nber.org/papers/w14657).

Bozo, Frédéric. *Two Strategies for Europe: De Gaulle, the United States and the Atlantic Alliance*. Lanham: Rowman and Littlefield, 2001.

Brands, Hal. Non-Proliferation and the Dynamics of the Middle Cold War: The Superpowers, the MLF and the NPT. *Cold War History* 7:3 (2007): 389–423.

Burr, William. "To "Keep the Genie Bottled Up": U.S. Diplomacy and Gas Centrifuge Technology, 1962–1972." *Journal of Cold War Studies*. In press.

Burr, William. "The "Labors of Atlas, Sisyphus, or Hercules"? US Gas-Centrifuge Policy and Diplomacy, 1954–1960." *International History Review* 37 (3) (2014): 431–457.

Burr, William. A Scheme of "Control": The United States and the Origins of the Nuclear Suppliers' Group, 1974–1976. *International History Review* 36:2 (2014): 252–276.

Burr, William, and Jeffrey T. Richelson. Whether to "Strangle the Baby in the Cradle": The United States and the Chinese Nuclear Program, 1960–1964. *International Security* 25:3 (2000–01): 55–99.

Butler, Stuart A. The Struggle for Power: Britain and Euratom, 1955–1963. *International History Review* 36:2 (2014): 324–341.

Carson, Cathryn. *Heisenberg in the Atomic Age: Science and the Public Sphere.* Cambridge University Press, 2010.

Chadeau, E., ed. *L'ambition technologique: naissance d'Ariane.* Paris: Rive Droite, 1995.

Chernus, Ira, *Eisenhower's Atoms for Peace* (Texas A&M University Press, 2002).

Connelly, Matthew. *Fatal Misconception: The Struggle to Control World Population.* Cambridge: Belknap, 2006.

Costigliola, Frank. *France and the United States.* New York: Twayne, 1992.

Creager, Angela. *Life Atomic: A History of Radioisotopes in Science and Medicine.* University of Chicago Press, 2013.

Cueto, Marcus. *Cold War, Deadly Fevers: Malaria Eradication in Mexico, 1955–1975.* Baltimore: Johns Hopkins University Press, 2007.

Cullather, Nick. Miracles of Modernization: The Green Revolution and the Apotheosis of Technology. *Diplomatic History* 28:2 (2004): 227–254.

Cullather, Nick. *The Hungry World: America's Cold War Battle Against Poverty in Asia.* Cambridge: Harvard University Press, 2010.

De Grazia, Victoria. *Irresistible Empire: America's Advance through 20th-Century Europe.* Cambridge: Belknap, 2005.

Doel, Ron E. Science in Black: Scientific Intelligence and Scientific Practice in the Cold War Era. In *Cold War Science and the Transatlantic Circulation of Knowledge*, ed. Jeroen Van Dongen. Leiden: Brill, 2015.

Dumoulin, Michel, Pierre Guillen, and Maurice Vaïsse, eds. *L'Énergie nucléaire en Europe: Des origins à Euratom.* Bern: Peter Lang, 1994.

Eckert, Michael. Primacy Doomed to Failure: Heisenberg's Role as Scientific Adviser for Nuclear Policy in the FRG. *Historical Studies in the Physical and Biological Sciences* 21:1 (1990): 29–58.

Eckert, Michael. Kernenergie und Wesintegration: Die Zähmung des westdeutschen Nuklearnazionalizmus. In *Vom Marshallplan zur EWG: Die Engliederung der Bundesrepublik Deutschland in die westliche Wëlt*, ed. Ludolf Herbst, Werner Bührer, and Hanna Sowade. Munich: Oldenbourg, 1990.

Edgerton, David E. H., The "White Heat" Revisited: The British Government and Technology in the 1960s. *Twentieth Century British History* 7:1 (1996): 53–82.

Edgerton, David. *Warfare State: Britain 1920–1970*. Cambridge University Press, 2006.

Edgerton, David. The Contradictions of Techno-nationalism and Techno-globalism: A Historical Perspective. *New Global Studies* 1:1 (2007). DOI: 10.2202/1940-0004.1013.

Elli, M. *Politica estera ed ingegneria nucleare: I rapporti del Regno Unito con l'Euratom, 1957–1963*. Milan: Unicopli, 2007.

Engerman, David C. American Knowledge and Global Power. *Diplomatic History* 31:4 (2007): 599–621.

Engerman, D. C., N. Gilman, M. E. Haefele, and M. E. Latham, eds. *Staging Growth: Modernization, Development and the Global Cold War*. Boston: University of Massachusetts Press, 2003.

Friedberg, Aaron L. *In the Shadow of the Garrison State: America's Anti-Statism and Its Cold War Grand Strategy*. Princeton University Press, 2000.

Frutkin, Arnold W. The United States Space Program and Its International Significance. *Annals of the American Academy of Political and Social Science* 366:July (1966): 89–98.

Fuhrmann, Matthew. *Atomic Assistance: How "Atoms for Peace" Programs Cause Nuclear Insecurity*. Ithaca: Cornell University Press, 2012.

Gallois, Pierre. *Stratégie de l'âge nucléaire*. Paris: Calmann-Lévy, 1960.

Gavin, Francis J. *Nuclear Statecraft: History and Strategy in America's Atomic Age*. Ithaca: Cornell University Press, 2012.

Gillon, Luc. L'approvisionnement en uranium. In *L'Énergie nucléaire en Europe: Des origins à Euratom*, ed. Michel Dumoulin, Pierre Guillen and Maurice Vaïsse. Bern: Peter Lang, 1994.

Goldschmidt, Bertrand. *The Atomic Complex: A Worldwide Political History of Nuclear Energy*. LaGrange Park: American Nuclear Society, 1982.

Granieri, Ronald J. Odd Man Out? The CDU-CSU, Ostpolitik, and the Atlantic Alliance. In *The Strained Alliance: US-European Relations from Nixon to Carter*, ed. Matthias Schulz and Thomas A. Schwartz. German Historical Institute and Cambridge University Press, 2010.

Greenhalgh, Susan. The Social Construction of Population Science: An Intellectual, Institutional and Political History of Twentieth-Century Demography. *Comparative Studies in Society and History* 38:1 (1996): 26–66.

Guillen, Pierre. La France et la négociation du traité d'Euratom. In *L'Énergie nucléaire en Europe: Des origins à Euratom*, ed. Michel Dumoulin, Pierre Guillen and Maurice Vaïsse. Bern: Peter Lang, 1994.

Hamblin, Jacob Darwin. The Nuclearization of Iran in the Seventies. *Diplomatic History* 38:5 (2014): 1114–1135.

Hecht, Gabrielle. *The Radiance of France: Nuclear Power and National Identity After World War II*. Cambridge: MIT Press, 2009.

Hecht, G., ed. *Entangled Geographies: Empire and Techno-politics in the Global Cold War*. Cambridge: MIT Press, 2011.

Hecht, Gabrielle. *Being Nuclear: Africans and the Global Uranium Trade*. Cambridge: MIT Press, 2012.

Hecht, Gabrielle, and Paul N. Edwards. *The Technopolitics of Cold War*. American Historical Association, 2007.

Helmreich, Jonathan D. The United States and the Formation of Euratom. *Diplomatic History* 15:3 (1991): 387–410.

Herken, Greg. "A Most Deadly Illusion": The Atomic Secret and American Nuclear Weapons Policy, 1945–1950. *Pacific Historical Review* 49:1 (1980): 51–76.

Hewlett, Richard G., and Jack M. Holl. *Atoms for Peace and War, 1953–1961: Eisenhower and the Atomic Energy Commission*. Berkeley: University of California Press, 1989.

Hitchcock, William I. *The Struggle for Europe: The Turbulent History of a Divided Continent, 1945 to the Present*. New York: Anchor Books, 2004.

Hogan, Michael J. *A Cross of Iron: Harry S. Truman and the Origins of the National Security State*. Cambridge University Press, 1998.

Holl, Jack M. The Peaceful Atom: Lore and Myth. In *Atoms for Peace: An Analysis after Thirty Years*, ed. Joseph F. Pilat, Robert E. Pendley and Charles K. Ebinger. Boulder: Westview.

Iriye, Akira. Transnational History. *Contemporary European History* 13:2 (2004): 211–222.

Johnston, Sean F. *The Neutron's Children: Nuclear Engineers and the Shaping of Identity*. Oxford University Press, 2012.

Judt, Tony. The Illusionist: Henry Kissinger and American Foreign Policy. In *Reappraisals: Reflections on the Forgotten Twentieth Century*, ed. Tony Judt. New York: Penguin, 2008.

Kaiser, Wolfram. *Using Europe, Abusing the Europeans*. New York: St. Martin's Press, 1996.

Kemp, R. Scott. "Gas Centrifuge Theory and Development: A Review of US Programs." *Science and Global Studies* 17 (2009): 1–19.

Kemp, R. Scott. Nonproliferation Strategy in the Centrifuge Age, PhD dissertation, Princeton University, 2010.

Kemp, R. Scott. "The End of Manhattan: How the Gas Centrifuge Changed the Quest for Nuclear Weapons." *Technology and Culture* 53 (2) (2012): 272–305.

Kevles, Daniel J. Principles and Politics in Federal R&D Policy, 1945–1990. An Appreciation of the Bush Report. In *Vannevar Bush, Science—The Endless Frontier, 40th Anniversary Edition*. Washington: National Science Foundation, 1990.

Killian, James R., *Sputniks, Scientists and Eisenhower: A Memoir of the First Special Assistant to the President for Science and Technology*. Cambridge: MIT Press, 1977.

Klein, Cristina. *Cold War Orientalism: Asia in the Middlebrow Imagination, 1945–1961*. Berkeley: University of California Press, 2003.

Knorr, Klaus E. Euratom and American Policy: A Conference Report. Center for International Studies, Princeton University, 1956.

Kohlrausch, Martin, and Helmuth Trischler. *Building Europe on Expertise: Innovators, Organizers, Networkers*. New York: Palgrave Macmillan, 2014.

Krige, John. "NATO and the Strengthening of Western Science in the Post-Sputnik Era." *Minerva* 38 (2000): 81–108.

Krige, John. *American Hegemony and the Postwar Reconstruction of Science in Europe*. Cambridge: MIT Press, 2006.

Krige, John. Atoms for Peace, Scientific Internationalism, and Scientific Intelligence. In *Global Power Knowledge: Science and Technology in International Affairs*, ed. John Krige and Kai-Henrik Barth (*Osiris* 21, 2006).

Krige, John. The Peaceful Atom as Political Weapon: Euratom and American Foreign Policy in the Late 1950s. *Historical Studies in the Natural Sciences* 38:1 (2008): 5–44.

Krige, John. Techno-Utopian Dreams, Techno-Political Realities. The Education of Desire for the Peaceful Atom. In *Utopia/Dystopia: Conditions of Historical Possibility*, ed. Michael D. Gordin, Helen Tilley, and Gyan Prakash. Princeton University Press, 2010.

Krige, John. Technological Leadership and American Soft Power. In *Soft Power and US Foreign Policy: Theoretical, Historical and Contemporary Perspectives*, ed. Inderjeet Parmar and Michael Cox. London: Routledge, 2010.

Krige, John. Building the Arsenal of Knowledge. *Centaurus* 52 (2010): 280–296.

Krige, John. The Proliferation Risks of Gas Centrifuge Enrichment at the Dawn of the NPT: Shedding Light on the Negotiating History. *Nonproliferation Review* 19:2 (2012): 219–227.

Krige, John. Hybrid Knowledge: The Transnational Coproduction of the Gas Centrifuge for Isotope Enrichment in the 1960s. *British Journal for the History of Science* 45:3 (2012): 337–357.

Krige, John. US Technological Superiority and the Special Nuclear Relationship: Contrasting British and US Policies for Controlling the Proliferation of Gas-Centrifuge Enrichment. *International History Review* 36:2 (2014): 230–251.

Krige, John. Diplomacy (Post 1945), Science and Technology. In *Oxford Encyclopedia of the History of American Science, Medicine and Technology*, volume I, ed. Hugh R. Slotten. Oxford University Press, 2014.

Krige, John. "Euratom and the IAEA: The Problem of Self-Inspection." *Cold War History* 15 (3) (2015): 341–352.

Krige, John, and Arturo Russo. A History of the European Space Agency, 1958–1987, volume I: The Story of ESRO and ELDO, 1958–1973. Report SP-1235, European Space Agency.

Krige, John, Angelina Long Callahan, and Ashok Maharaj. *NASA in the World: Fifty Years of International Collaboration in Space*. New York: Palgrave Macmillan, 2013.

Krige, J., and H. Rausch, eds. *American Foundations and the Coproduction of World Order in the Twentieth Century*. Göttingen: Vandenhoeck & Ruprecht, 2012.

Kroenig, Matthew. *Exporting the Bomb: Technology Transfer and the Spread of Nuclear Weapons*. Ithaca: Cornell University Press, 2010.

Küntzel, Matthias. *Bonn and the Bomb: German Politics and the Nuclear Option*. London: Pluto, 1995.

LaFeber, Walter. Technology and US Foreign Relations. *Diplomatic History* 24:1 (2000), 1–19.

Latham, Michael E. *The Right Kind of Revolution: Modernization, Development, and US Foreign Policy from the Cold War to the Present*. Ithaca: Cornell University Press, 2010.

Leffler, Melvyn. *For the Soul of Mankind: The United States, the Soviet Union and the Cold War*. New York: Hill and Wang, 2007.

Logsdon, John M., Learning from the Leader. The Early Years of US-Japanese Cooperation in Space. Report, Space Policy Institute, Elliot School of International Affairs, George Washington University, undated.

Logsdon, John M. *John F Kennedy and the Race to the Moon*. New York: Palgrave Macmillan, 2012.

Ludlow, Piers. From Words to Actions: Reinterpreting de Gaulle's European Policy. In *Globalizing de Gaulle: International Perspectives on French Foreign Policies, 1958–1969*, ed. Christian Nuenlist, Anna Locher and Garret Martin. Plymouth: Lexington Books, 2010.

Ludlow, N. Piers. European Integration and the Cold War. In *The Cambridge History of the Cold War*, volume 2: *Crises and Détente*, ed. Melvyn P. Leffler and Odd Arne Westad. Cambridge University Press, 2010.

Lundestad, Geir. *"Empire" by Invitation: The United States and European Integration, 1945–1997*. Oxford University Press, 1998.

Lundestad, Geir. *The United States and Western Europe Since 1945*. Oxford University Press, 2003.

Maddock, Shane. *Nuclear Apartheid: The Quest for American Supremacy From World War II to the Present*. Chapel Hill: University of North Carolina Press, 2010.

Mallard, Grégoire. Atomic Confederacy: Europe's Quest for Nuclear Weapons and the New World Order. PhD thesis, Princeton University, 2008.

Mallard, Grégoire. L'Europe Puissance Nucléaire, Cet Obscure Objet du Désir. *Critique internationale* 1:42 (2009): 141–163.

Mallard, Grégoire. *Fallout: Nuclear Diplomacy in an Age of Global Fracture*. University of Chicago Press, 2014.

Mallard, Grégoire, and Andrew Lakoff. How Claims to Know the Future Are Used to Understand the Present. In *Social Science in the Making*, ed. Charles Camic, Michèle Lamont, and Neil Gross. University of Chicago Press, 2011.

Manela, Erez. A Pox on Your Narrative: Writing Disease Control into Cold War History. *Diplomatic History* 34:2 (2010): 299–323.

McDougall, Walter A. Space Age Europe: Gaullism, Euro-Gaullism and the European Dilemma. *Technology and Culture* 26:2 (1985): 179–203.

McDougall, Walter A. *The Heavens and the Earth: A Political History of the Space Age*. Baltimore: Johns Hopkins University Press, 1997.

McGrath, Patrick J. *Scientists, Business, and the State, 1890–1960*. Chapel Hill: University of North Carolina Press, 2002.

MacKenzie, Donald. *Inventing Accuracy: Historical Sociology of Nuclear Missile Guidance*. Cambridge: MIT Press, 1993.

Medhurst, Martin J. Atoms for Peace and Nuclear Hegemony: The Rhetorical Structure of a Cold War Campaign. *Armed Forces and Society* 23 (1997): 571–593.

Melandri, Pierre. Aux Origines de la Coopération Nucléaire Franco-Américaine. In *Le France et l'Atome: Etudes de l'histoire nucléaire*, ed. Maurice Vaïsse. Brussels: Bruylan, 1994.

Meyer, Stephen M. *The Dynamics of Nuclear Proliferation*. University of Chicago Press, 1984.

Milward, Alan S. *The United Kingdom and the European Community: The Rise and Fall of a National Strategy, 1945–1963*, volume 1. London: Routledge, 2013.

Morland, Howard. Born Secret. *Cardozo Law Review* 26:4 (2005): 1401–1408.

Neufeld, Michael. *Von Braun: Dreamer of Space, Engineer of War*. New York: Vintage, 2008.

Newhouse, John. *De Gaulle and the Anglo-Saxons*. New York: Viking, 1970.

Nicolaidis, L., R. Ockrent, and W. Harpham. *Possibilities of Action in the Field of Nuclear Energy*. Paris: Organization for European Economic Cooperation, 1956.

Nieburg, H. L. Euratom: A Study in Coalition Politics. *World Politics* 15 (1963): 597–622.

Nuenlist, C., A. Locher, and G. Martin, eds. *Globalizing de Gaulle: International Perspectives on French Foreign Policies, 1958–1969*. Plymouth: Lexington Books, 2010.

Nye, Joseph S., Jr. Soft Power. *Foreign Policy* 80:fall (1990): 153–172.

Nye, Joseph S., Jr. *Soft Power: The Means to Success in World Politics*. New York: Public Affairs, 2004.

O'Driscoll, Mervyn. Missing the Nuclear Boat: British Policy and French Nuclear Ambitions during the Euratom Foundation Negotiations, 1955-56. *Diplomacy and Statecraft* 1 (1998): 135–162.

Oldenziel, Ruth, and Karin Zachmann , eds. *Cold War Kitchen: Americanization, Technology and European Users* Cambridge: MIT Press, 2011.

Oppenheimer, J. Robert. Testimony. In *The Great Inquiry: Testimony at AEC Hearings. Bulletin of the Atomic Scientists* 5:8–9 (1949), 221–254.

Osgood, Kenneth. *Total Cold War: Eisenhower's Secret Propaganda Battle at Home and Abroad.* Lawrence: University Press of Kansas, 2006.

Parr, Helen. *Britain's Policy Towards the European Community: Harold Wilson and Britain's World Role, 1964–1967.* London: Routledge, 2006.

Paxton, Robert O., and Nicholas Wahl, eds., *De Gaulle and the United States, A Centennial Reappraisal.* Oxford: Berg, 1994.

Pilat, Joseph F., Robert E. Pendley, and Chales K. Ebinger, eds. *Atoms for Peace: An Analysis after Thirty Years.* Boulder: Westview, 1985.

Popp, Roland. Introduction: Global Order, Cooperation between the Superpowers, and Alliance Politics in the Making of the Nuclear Non-Proliferation Regime. *International History Review* 36:2 (2014): 195–209.

Proceedings of the International Conference on the Peaceful Uses of Atomic Energy, Held in Geneva, 8 August–20 August 1955, volume 16. New York: United Nations, 1956.

Quist, Arvin S. Security Classification of Information, volume I. Report ORCA-12, Oak Ridge Classification Associates, LLC, 2002.

Reinke, Niklas. *The History of German Space Policy: Ideas, Influences and Interdependence, 1923–2002.* Paris: Beauchesne, 2007.

Rey, Marie-Pierre. De Gaulle, French Diplomacy and Franco-Soviet Relations as Seen from Moscow. In *Globalizing de Gaulle: International Perspectives on French Foreign Policies, 1958–1969*, ed. Christian Nuenlist, Anna Locher, and Garret Martin. Plymouth: Lexington Books, 2010.

Rohde, Joy. *Armed with Expertise: The Militarization of American Social Research in the Cold War.* Ithaca: Cornell University Press, 2013.

Roland, Alex. *The Military-Industrial Complex.* AHA-SHOT Publications, 2001.

Sagan, Scott D. Why do States Build Nuclear Weapons? Three Models in Search of a Bomb. *International Security* 21:3 (1996–97): 54–86.

Saunders, Frances S. *Who Paid the Piper? The CIA and the Cultural Cold War.* London: Granta, 2000.

Schrafstetter, Susanna. The Long Shadow of the Past: History, Memory and the Debate over West Germany's Nuclear Status, 1954–69. *History & Memory* 16:1 (2004): 118–145.

Schrafstetter, Susan, and Stephen Twigge. Spinning into Europe: Britain, West Germany and the Netherlands—Uranium Enrichment and the Development of the Gas Centrifuge, 1964–1970. *Contemporary European History* 11 (2002): 253–272.

Schrafstetter, Susan, and Stephen Twigge, *Avoiding Armageddon: Europe, the United States, and the Struggle for Nuclear Non-Proliferation, 1945–1970*. Westport: Praeger, 2004.

Scheinman, Lawrence. *Atomic Energy Policy in France Under The Fourth Republic*. Princeton University Press, 1965.

Scheinman, Lawrence. *The International Atomic Energy Agency and World Order*. Washington: Resources for the Future, 1987.

Schwabe, Klaus. The United States and European Integration. In *Western Europe and Germany: The Beginnings of European Integration. 1945–1960*, ed. Clemens Wurm. Oxford: Berg, 1995.

Schwartz, Hans-Peter, *The Statesman, 1952–1967*. Providence: Berghahn Books, 1997.

Schwartz, Thomas Alan, *Lyndon Johnson and Europe: In the Shadow of Vietnam*. Cambridge: Harvard University Press, 2003.

Scott-Smith, Giles, *Networks of Empire: The US State Department's Foreign Leader Program in the Netherlands, France and Britain 1950–1970*. Brussels: Peter Lang, 2008.

Seaborg, Glenn T. *Stemming the Tide: Arms Control in the Johnson Years*. Plymouth: Lexington Books, 1987.

Secord, James A. *Knowledge in Transit. Isis* 95 (2004): 654–672.

Servan-Schreiber, Jean-Jacques, *Le défi américain*. Paris: Denoël, 1967.

Skogmar, Gunnar, *The United States and the Nuclear Dimension of European Integration*. London: Palgrave Macmillan, 2004.

Sokolski, Henry D. *Best of Intentions: America's Campaign Against Strategic Weapons Proliferation*. Westport: Praeger, 2001.

Solovey, Mark, and Hamilton Cravens. *Cold War Social Science: Knowledge Production, Liberal Democracy and Human Nature*. New York: Palgrave Macmillan, 2014.

Soutou, Georges-Henri. La politique nucléaire de Pierre Mendès France. In *La France et l'atome: Études d'histoire nucléaire*, ed. Maurice Vaîsse. Brussels: Bruylant, 1994.

Soutou, Georges-Henri. Le Président Pompidou et les relations entre les Etats-Unis et l'Europe. *Journal of European Integration History* 6:2 (2000): 111–146.

Soutou, Georges-Henri. Georges Pompidou and US-European Relations. In *Between Empire and Alliance: America and Europe During the Cold War*, ed. Marc Trachtenberg. New York: Rowman and Littlefield, 2003.

Suri, Jeremi. Henry Kissinger and the Geopolitics of Globalization. In *The Shock of the Global: The 1970s in Perspective*, ed. Niall Ferguson, Charles Maier, Erez Manela, and Daniel Sargent. Cambridge: Belknap, 2011.

Swango, Dane. The United States and the Role of Nuclear Co-operation and Assistance in the Design of the Non-Proliferation Treaty. *International History Review* 36:2 (2014): 210–229.

Trachtenberg, Marc. *A Constructed Peace: The Making of the European Settlement, 1945–1963*. Princeton University Press, 1999.

Trachtenberg, Marc. The French Factor in US Foreign Policy during the Nixon-Pompidou Period, 1969–1974. *Journal of Cold War Studies* 13:1 (2011): 4–59.

Trachtenberg, Marc. The State of International History: Where We've Been, Where We Are, and Where We're Going. H-Diplo Daily Digest, July 4, 2014 (https://networks.h-net.org/node/28443/discussions/33922/h-diplo-state-field-essays.

Trischler, Helmuth, and Hans Weinberger. Engineering Europe: Big Technologies and Military Systems in the Making of Twentieth Century Europe. *History and Technology* 25:1 (2005): 49–83.

Turchetti, Simone. A Most Active Customer: How the US Administration Helped the Italian Atomic Energy Project to "De-Develop." *Historical Studies in the Natural Sciences* 44:5 (2014): 470–502.

Twigge, Stephen. A Baffling Experience: Technology Transfer, Anglo-American Nuclear Relations, and the Development of the Gas Centrifuge, 1964–1970. *History and Technology* 19 (2003): 151–163.

Twigge, Stephen.R. *The Early Development of Guided Weapons in the United Kingdom, 1940–1960*. London: Routledge, 1993.

Twigge, Stephen. A Baffling Experience: Technology Transfer, Anglo-American Nuclear Relations, and the Development of the Gas Centrifuge, 1964–1970. *History and Technology* 19 (2003): 151–163.

Twigge, Stephen. The Third Nuclear Power. Britain and the Beginnings of the IAEA. Paper presented at conference United Atoms in a Divided World. The Early History of the International Atomic Energy Agency, Vienna, 2012.

Ullman, Richard. The Covert French Connection. *Foreign Policy* 75:summer (1983): 3–33.

Vaîsse, Maurice. Les 'relations spéciales' franco-américaines au temps de Richard Nixon et Georges Pompidou. *Relations Internationales* 119 (2004): 345–362.

Vaïsse, Maurice. *La Puissance ou l'Influence? La France dans le monde depuis 1958*. Paris: Fayard, 2009.

Vaïsse, Maurice. *La Grandeur politique étrangére du général de Gaulle 1958–1969*. Paris: Fayard, 1998.

Walker, William, and Måns Lönnroth. Arms Control or Anarchy. In *Atoms for Peace: An Analysis after Thirty Years*, ed. Joseph F. Pilat, Robert E. Pendley, and Charles K. Ebinger. Boulder: Westview.

Weart, Spencer R. *Nuclear Fear: A History of Images*. Cambridge: Harvard University Press, 1998.

Weiner, Charles. A New Site for the Seminar: the Refugees and American Physics in the Thirties. In *The Intellectual Migration: Europe and America, 1930–1960*, ed. D. Fleming and B. Bailyn. Cambridge: Belknap, 1969.

Westad, Odd Arne. The New International History of the Cold War: Three (Possible) Paradigms. *Diplomatic History* 24:4 (2000): 551–565.

Westad, Odd Arne. *The Global Cold War: Third World Interventions and the Making of Our Times*. Cambridge University Press, 2007.

Winand, Pascaline. *Kennedy, Eisenhower, and the United States of Europe*. St. Martin's Press, 1993.

Wolfe, Audra, *Competing with the Soviets: Science, Technology, and the State in Cold War America*. Baltimore: Johns Hopkins University Press, 2014.

Wood, Houston G., Alexander Glaser, and R. Scott Kemp. The Gas Centrifuge and Nuclear Weapons Proliferation. *Physics Today* 61:9 (2008): 40–45.

Young, John W. *Britain and European Unity, 1945–1992*. New York: St. Martin's Press, 1993.

Zimmerman, Hubert. *Money and Security: Troops Monetary Policy, and West Germany's Relations with the United States and Britain, 1950–1971*. Cambridge University Press, 2002.

Zuckerman, Solly, *Monkeys, Men and Missiles: An Autobiography, 1964–1988*. London: Collins, 1988.

Index